M000237154

Shedding Light on Electricity Utilities in the Middle East and North Africa

DIRECTIONS IN DEVELOPMENT
Energy and Mining

Shedding Light on Electricity Utilities in the Middle East and North Africa

Insights from a Performance Diagnostic

Daniel Camos, Robert Bacon, Antonio Estache, and Mohamad M. Hamid

LIBRARY OF
CONGRESS
SURPLUS
DUPLICATE

© 2018 International Bank for Reconstruction and Development / The World Bank
1818 H Street NW, Washington, DC 20433
Telephone: 202-473-1000; Internet: www.worldbank.org

Some rights reserved

1 2 3 4 21 20 19 18

This work is a product of the staff of The World Bank with external contributions. The findings, interpretations, and conclusions expressed in this work do not necessarily reflect the views of The World Bank, its Board of Executive Directors, or the governments they represent. The World Bank does not guarantee the accuracy of the data included in this work. The boundaries, colors, denominations, and other information shown on any map in this work do not imply any judgment on the part of The World Bank concerning the legal status of any territory or the endorsement or acceptance of such boundaries.

Nothing herein shall constitute or be considered to be a limitation upon or waiver of the privileges and immunities of The World Bank, all of which are specifically reserved.

Rights and Permissions

This work is available under the Creative Commons Attribution 3.0 IGO license (CC BY 3.0 IGO) http://creativecommons.org/licenses/by/3.0/igo. Under the Creative Commons Attribution license, you are free to copy, distribute, transmit, and adapt this work, including for commercial purposes, under the following conditions:

Attribution—Please cite the work as follows: Camos, Daniel, Robert Bacon, Antonio Estache, and Mohamad M. Hamid. 2018. *Shedding Light on Electricity Utilities in the Middle East and North Africa: Insights from a Performance Diagnostic*. Directions in Development. Washington, DC: World Bank. doi:10.1596/978-1-4648-1182-1. License: Creative Commons Attribution CC BY 3.0 IGO

Translations—If you create a translation of this work, please add the following disclaimer along with the attribution: *This translation was not created by The World Bank and should not be considered an official World Bank translation. The World Bank shall not be liable for any content or error in this translation.*

Adaptations—If you create an adaptation of this work, please add the following disclaimer along with the attribution: *This is an adaptation of an original work by The World Bank. Views and opinions expressed in the adaptation are the sole responsibility of the author or authors of the adaptation and are not endorsed by The World Bank.*

Third-party content—The World Bank does not necessarily own each component of the content contained within the work. The World Bank therefore does not warrant that the use of any third-party-owned individual component or part contained in the work will not infringe on the rights of those third parties. The risk of claims resulting from such infringement rests solely with you. If you wish to re-use a component of the work, it is your responsibility to determine whether permission is needed for that re-use and to obtain permission from the copyright owner. Examples of components can include, but are not limited to, tables, figures, or images.

All queries on rights and licenses should be addressed to World Bank Publications, The World Bank Group, 1818 H Street NW, Washington, DC 20433, USA; e-mail: pubrights@worldbank.org.

ISBN (paper): 978-1-4648-1182-1
ISBN (electronic): 978-1-4648-1183-8
DOI: 10/1596/978-1-4648-1182-1

Cover photo: © Anna Om / Shutterstock. Night cityscape in Lebanon. Used with the permission. Further permission required for reuse.
Cover design: Debra Naylor, Naylor Design, Washington, DC

Library of Congress Cataloging-in-Publication Data have been requested.

Contents

Tables

Acknowledgments

This study was written by Daniel Camos, Robert Bacon, Antonio Estache, and Mohamad Mahgoub Hamid. The core team included Bipul Singh, Adnan Sirajee, and Mark Njore. It was carried out under the guidance of Erik Fernstrom and Charles Cormier. Vivien Foster provided substantial support and feedback throughout the exercise. The team gratefully acknowledges peer review comments received from Luis Andres, Sudeshna Banerjee, Vivien Foster, Victor Loksha, Marcelino Madrigal, Elvira Morella, and Sameer Shukla. We are grateful to Franck Bousquet and to Jonathan Walters, who provided guidance and advice, the latter in the concept stage and the former in the final stage.

Electricity utilities, line ministries, and regulators are gratefully acknowledged for sharing data. This study also benefited from numerous exchanges with the Arab Union of Electricity, the Arab Electricity Regulatory Forum, and the Regional Center for Renewable Energy and Energy Efficiency, which also provided support in gathering data from publicly available annual reports and utility financial statements.

We are grateful to the following World Bank colleagues who provided advice and helped in data collection and validation: Waleed Alsuraih, Husam Beides, Roger Coma, Ferhat Esen, Mohab Hallouda, Ashish Khanna, Fanny Missfeldt-Ringius, Alejandro Moreno, Tara Shirvani, Simon Stolp, Manaf Touati, Chris Trimble, and Jianping Zhao.

The data questionnaire was developed by Richard Schlirf, with support from Daniel Camos, Bipul Singh, Sudeshna Banerjee, and Marcelino Madrigal, all under the guidance of Vivien Foster. The data collection process was led by Daniel Camos and Bipul Singh. Jorge Sneij provided invaluable advice on information architecture and managing heavy datasets.

The following people provided excellent support during data collection: Manaf Touati and Badr El Ahrari for Algeria; Hafez El Salmawy, Fatma Mostafa, and Bipul Singh for the Arab Republic of Egypt; Georges Dib and Bipul Singh for Bahrain; Aboubakar Hassan and Roger Coma for Djibouti; Simon Stolp and Adnan Sirajee for Iraq; Salah Tayeh and Usaimah Khalifeh for Jordan; Georges Zammar for Lebanon; Manaf Touati, Tayeb Amegroud, and Badr El Ahrari for Morocco; Hassan Taqi and Zahra Al Obaidani for Oman; Ahmad Esmaeel Al Mutawkel for the Republic of Yemen; Adnan Sirajee for

Qatar; Mansour Helal Al-Anazi and Debasish Ghosh for Saudi Arabia; Ezzedine Khalfallah for Tunisia; and Reem Muhsin Yusuf for the West Bank.

Alberto Cena and Tayeb Amegroud provided substantial support in verifying and interpreting the data.

The country case studies were led by Bipul Singh and Fatma Mostafa (Egypt), Imad Nejdawi and Mohamad Mahgoub Hamid (Jordan), Tayeb Amegroud and Richard Schlirf (Morocco), and Kirstin Morrison and Adnan Sirajee (Oman).

Steven Kennedy provided excellent editing support, and Jewel McFadden effectively managed publication aspects.

The financial and technical support by the Energy Sector Management Assistance Program (ESMAP) is gratefully acknowledged. ESMAP—a global knowledge and technical assistance program administered by the World Bank—assists low- and middle-income countries to increase their know-how and institutional capacity to achieve environmentally sustainable energy solutions for poverty reduction and economic growth. ESMAP is funded by Australia, Austria, Denmark, the European Commission, Finland, France, Germany, Iceland, Italy, Japan, Lithuania, Luxembourg, the Netherlands, Norway, the Rockefeller Foundation, Sweden, Switzerland, the United Kingdom, and the World Bank.

About the Authors

Daniel Camos is a senior infrastructure economist at the World Bank, where he has worked in the energy and water global practices leading both operations and analytical work. Previously, he worked for the European Commission, the United Nations, and nongovernmental organizations. He currently works in the MENA region and has previous experience in Latin America and the Caribbean and in Sub-Saharan Africa. He has a training in economics and engineering, including a PhD in economics from the Paris School of Economics and the Université libre de Bruxelles; an MPA in international development from Harvard Kennedy School; and an industrial engineering degree from the Polytechnic University of Catalonia.

Robert Bacon is a specialist in the economics of the energy sector. In recent years, he has been a consultant with the World Bank and the African Development Bank. Before that, he was the manager of the Oil and Gas Policy Division of the World Bank. Previously, for 30 years he was on the faculty of economics at the University of Oxford and a visiting fellow at the Oxford Institute for Energy Studies. His work on the energy sector includes studies on the assessment of progress and problems of reform of power sectors in low- and middle-income countries, econometric analyses of pricing and demand for petroleum products, and evaluation of the generation of employment throughout an economy by the actions of the energy sector.

Antonio Estache is professor of economics at the Université libre de Bruxelles where he holds the Bernard Vanommeslaghe Chair aimed at increasing awareness throughout Europe about the importance of regulation and competition issues in public service industries. He is also a member of the European Center for Advanced Research in Economics and Statistics in Brussels. Previously, he spent 25 years at the World Bank, where he worked on infrastructure restructuring, procurement, regulation, and public sector and tax reform. He has published extensively on these topics and continues to work as a policy adviser to international organizations, governments, and parliaments around the world.

Mohamad Mahgoub Hamid has an engineering background in mechanics and energetics and has recently been working as a consultant for the World Bank. His experience in the MENA region includes research and project coordination at the Regional Centre for Renewable Energy and Energy Efficiency in Cairo. As an energy policy analyst, he worked on projects with the League of Arab States and several United Nations agencies. Mohamad has followed trainings in several energy-related fields, including an energy statistics course at the International Energy Agency in Paris. He has a masters in engineering and a masters in management from the University of Aix-Marseille.

Abbreviations

AER	Authority for Electricity Regulation
AICD	Africa Infrastructure Country Diagnostic
ANRE	Agence Nationale de Régulation de l'Electricité
BOOT	build-own-operate-transfer
CAIDI	Customer Average Interruption Duration Index
CAPEX	capital expenditure
CREG	Commission de Régulation de l'Electricité et du Gaz (Electricity and Gas Regulatory Commission)
DRSC	Direction des Régies et Services Concédés
DSP	distribution service provider
DU	distribution utility
ECRA	Electricity and Cogeneration Regulatory Authority
EEP	electric energy production
Egypt ERA	Egypt Electric Utility and Consumer Protection Regulatory Agency
EHV	extra high voltage
EMRC	Energy and Minerals Regulatory Commission
FTE	full-time equivalent
GCC	Gulf Cooperation Council
GDP	gross domestic product
GNI	gross national income
GU	generation utility
HFO	heavy fuel oil
HIC	high-income country
HV	high voltage
IAS	international accounting standards
IEA	International Energy Agency
IFRS	international financial reporting standards
IPP	independent power producer
IRENA	International Renewable Energy Agency

ISO	independent system operator
IWPP	independent water and power producer
JREEF	Jordan Renewable Energy and Energy Efficiency Fund
LAC	Latin America and the Caribbean
LMIC	low- and middle-income country
MASEN	Moroccan Agency for Solar Energy
MEMDD	Ministère de l'Enérgie, des Mines, et du Développement Durable (Ministry of Energy, Mines, and Sustainable Development)
MEMR	Ministry of Energy and Mineral Resources
MENA	Middle East and North Africa
MHEW	Ministry of Housing, Electricity and Water
MIS	main integrated system
MoERE	Ministry of Electricity and Renewable Energy
NREA	New and Renewable Energy Authority
OEB	Ontario Energy Board
OECD	Organisation for Economic Co-operation and Development
OFGEM	Office of Gas and Electricity Markets
OPEX	operational expenses
PERC	Palestinian Electricity Regulatory Council
PERG	Programme d'Electrification Rurale Global (Global Rural Electrification Program)
PHES	pumped hydroelectric energy storage
PPA	power purchase agreement
PPP	purchasing power parity
QFD	quasi-fiscal deficit
RE	renewable energy
RFP	request for proposal
RISE	Readiness for Investment in Sustainable Energy
ROA	return on assets
ROE	return on equity
SAIDI	System Average Interruption Duration Index
SAIFI	System Average Interruption Frequency Index
SAOG	Société Anonyme Omanaise Générale (Omani Public Limited Company)
SCADA	supervisory control and data acquisition
SPC	Services Permanents de Contrôle (Local Monitoring Units)
TL	transmission lines
TPA	third-party access
TSO	transmission system operator

TU	transmission utility
UMIC	upper-middle-income country
VIU	vertically integrated utility

All dollar amounts are U.S. dollars unless otherwise indicated.

Introduction

The Region's Electricity Challenge

The electricity sector in the Middle East and North Africa (MENA) is in the grip of an apparent paradox. Although the region continues to hold the world's largest oil and gas reserves and has been able to maintain electricity access rates of close to 100 percent in most of its economies, it may not be in a position to cater to the future electricity needs of its fast-growing population and their business activities. Primary energy demand in the region is expected to continue to rise at an annual rate of 1.9 percent through 2035, requiring a significant increase in generating capacity. Investments have not been rising fast enough to meet that requirement.

The annual electricity investments needed to keep up with demand have been estimated at about 3 percent of the region's projected gross domestic product (GDP) (Ianchovichina and others 2012). However, in most of the economies of the region, the ability to make those investments has been limited by fiscal constraints. The region's 2015 fiscal deficits averaged 9.3 percent of GDP, and the economies with the largest deficits were also those where electricity is most heavily subsidized. It seems unavoidable that, as economies adjust to their fiscal situation, they will continue to cut financing for the sector. To bridge the widening financing gap, the electricity sector must find its own financing sources, and it must do so quickly to keep pace with demand.

This work demonstrates that the solution is readily available: by improving the management and performance of the region's utilities, more than enough resources could be freed up to make the investments needed to meet demand and operate at lower cost. These management and policy changes would make the production and consumption of electricity more affordable for the region's taxpayers and could even make it more affordable for the poorest. They would also ease the transition toward renewable energy sources, reducing the dependency on imports for some economies and, for the economies that export oil and gas, extending the asset life of their nonrenewable resources.

The essence of the solution is not surprising. It involves cutting costs and improving revenue. But the report provides detailed evidence of the size of the potential gain. In short, efficiency improvements could generate financing equal to twice the sector's investment needs. That said, the optimal mix of cost-cutting and revenue-enhancing solutions is economy-specific, since cost and revenue-efficiency margins vary substantially across the region. For that reason, wherever several utilities share the responsibility to produce, transmit, and distribute electricity within a given economy, the analysis and the evidence identify the major cost drivers and the sources of revenue losses at the utility level.

The New MENA Electricity Database

This quantitative assessment of electricity utilities' performance has four main goals:

- To provide a recent, detailed **snapshot of technical and operational, commercial, and financial indicators** for a large sample of electricity utilities in the MENA region, based on a major effort to collect original data for the region
- To use these data to estimate the quasi-fiscal deficit (QFD) of the power sector in the economies of the region, and to determine what proportion of the deficit can be attributed to underpricing (setting tariffs below costs), collection losses (failure to bill or collect revenues due to the utility), transmission and distribution losses, and overstaffing (employing more labor than an efficient utility of the same size and characteristics would do)
- To assess the utilities' **relative performance** on a wide variety of indicators in MENA and beyond, as well as the **scope for improvements of MENA electricity utilities**, both at the utility and economy levels
- To **assess the relevance of key factors on operators' performance**—that is, the degree to which performance is affected by (a) vertical integration; (b) utility size; (c) utility ownership; (d) the presence or absence of a regulator; and (e) the level of development of a given economy.
- To distill useful lessons from four country case studies for the region to improve the performance of electricity utilities.

To provide answers to these questions, we surveyed the power utilities in the region and established the MENA Electricity Database (box I.1). Before this survey, information on the region's power sector was very uneven. The database thus forms a valuable public resource for policy makers as they reconcile the multiple dimensions of utility management performance with key policy concerns at the sector level. A limitation of the analysis is that the database's baseline is 2013, and the power sectors of some MENA economies have changed considerably since then.

The target audiences are **managers of electricity utilities, regulators, policy makers**, and other stakeholders (including members of civil society) concerned with the performance of specific utilities. The analysis is likely to be useful both at the sector level, since it highlights directions in which the sector may want to evolve in the region and in specific economies, and at the macroeconomic level,

Box I.1 The MENA Electricity Database

This study is based on collection and analysis of primary data on 36 performance indicators in the Middle East and North Africa (MENA) Electricity Database. It covers **67 electricity utilities in 14 economies** of the region: Algeria, the Arab Republic of Egypt, Bahrain, Djibouti, Iraq, Jordan, Lebanon, Morocco, Oman, Qatar, the Republic of Yemen, Saudi Arabia, Tunisia, and the West Bank.[a] It also relies on a sample of comparable non-MENA economies.

The data were collected by means of a standardized survey completed by utilities and regulatory agencies, covering indicators of technical, commercial, and financial performance. In some economies, the data were collected with support from local consultants or the public authorities. For the non-MENA economies, the data were collected from publicly available international databases. The sample of MENA operators comprises 12 vertically integrated utilities (VIUs), 29 distribution utilities (DUs), 23 generation utilities (GUs), and 3 transmission utilities (TUs). Data were collected from 2009 to 2013, with 2013 as the base year. Although the database contains much partial information, it also contains 945 base-year entries validated across 14 MENA economies and 3,832 entries for the period 2009–13.

Source: World Bank compilation.
a. Not included in the study are Libya, the Syrian Arab Republic, and the Islamic Republic of Iran. The utilities analyzed in this study are listed in appendix B.

since it highlights the main drivers of the fiscal costs of the sector. At the utility level, the data (where they are detailed enough) allow managers and regulators to evaluate performance features, which can them weigh the trade-offs involved in making utilities more cost-effective and client-oriented. For regulators and the other stakeholders concerned with the need to improve governance of the sector, the overall analysis highlights significant information gaps. Without data, poor management and poor policy decisions are unlikely to be addressed, imposing a significant cost on users and taxpayers.

The quality of the available data is also important. As a preliminary quality control measure, we asked utilities or economies to provide information on their accounting practices. First, we asked utilities about their adoption of (and compliance with) international accounting standards (IAS) or international financial reporting standards (IFRS): 60 percent responded positively, 10 percent negatively, and 30 percent did not respond. Second, we asked utilities whether they relied on cost-accounting systems; only one-third answered affirmatively—the other two-thirds were split between a negative answer and a nonresponse. Finally, the survey asked utilities if they relied on the supervisory control and data acquisition (SCADA) system of software and hardware elements to control processes locally or at remote locations or to monitor, gather, and process real-time data. Again, only one-third responded positively; the other two-thirds were split between a negative answer and a nonresponse. In sum, the quality of part of the available data—particularly that related to financial indicators—may be compromised by accounting practices.

The Structure and Content of the Report

The report is divided into two parts and several appendices. **Part I** (chapters 1–5) focuses on the region. **Part II** (chapters 6–10) consists of four country studies (Arab Republic of Egypt, Jordan, Morocco, and Oman) and a synopsis of all four. A short conclusion evokes the main themes and lessons from the entire report. Across the report, information at the utility level drawn from the MENA Electricity Database forms the basis of the analysis.

Chapter 1 calculates the QFD (or hidden costs) of the power sector in each of the 14 MENA economies studied, a first attempt to quantify the hidden costs of power sector inefficiencies in the region. QFDs are presented at the economy level and at the utility level. The hidden costs of financial, technical, commercial, and labor-related inefficiencies contribute to the already delicate fiscal situation of most economies in the MENA region and cause financial strains when they accumulate over several years. The QFD (or hidden-cost) approach has been used in numerous analyses as a powerful tool to communicate with policy makers. It also has been applied to other infrastructure sectors, notably water.[1]

The QFD was computed for 28 utilities, of which 11 are VIUs and 17 DUs. A limitation of this exercise was that it was not possible to compute the QFD for GUs and TUs, for lack of data on the price at which they sell electricity (a generation utility might sell to a TU or to a single buyer or VIU, depending on the structure of the market in the economy in which it operates).

Chapter 2 provides a snapshot of key performance indicators of MENA power utilities for which international comparisons are possible. These comparisons are made between the 14 MENA economies as well as with countries outside the region for which data were readily available. The MENA data are taken from the MENA Electricity Database. Comparisons are made for 14 technical, financial, and commercial indicators to highlight possible differences in performance among regions. Within MENA, further comparisons are made between utilities to highlight strong and weak performers for the indicator in question.

Ideally, comparisons for every indicator would be based on the same set of utilities within the region and on the same countries or utilities from outside it. However, this ideal is not yet attainable. The database has varying coverage for the 36 indicators included in the survey, for two reasons. First, certain indicators are relevant only to certain types of utilities. Second, many utilities did not report data on certain indicators, even when relevant: for example, only 46 of the 67 utilities surveyed reported data on their return on assets.

Chapter 3 examines performance indicators over time. Where governments have introduced power sector reform, policy makers should examine the reform's effects based on certain indicators. Changes should be expected to be gradual rather than sudden. To construct the MENA Electricity Database, we asked utilities to provide information for 2009–14. Because of the number of trend calculations to be made and the brevity of the data series, we decided to construct aggregates across utilities, indicator by indicator, and to carry out trend analysis on these aggregates for the few years of data available. Data were further disaggregated by utility type (distribution, generation, and vertically integrated) to check whether they revealed different trends.

Chapter 4 considers the relative overall performance of utilities within the MENA region when more than one indicator is considered. Understanding the reasons behind a strong overall performance can suggest policies that could be applied elsewhere, just as understanding poor performances can suggest ways to ameliorate the problem. The variability of performance across indicators suggests that some form of average performance measure is required if utilities are to be seen in a context of overall strengths and weaknesses.

The challenge in creating a multi-indicator approach is to combine indicators measured in very different units. This is done by the use of the "average rank score": the average of the ranked positions among utilities over a number of indicators.

Average rank scores were calculated for each of the 17 DUs for which we had data on five indicators, 13 GUs for which we had data on three indicators, and 8 VIUs for which we had data on three indicators. This exercise made it possible to identify the best- and worst-performing utilities within the groups analyzed.

Chapter 5 investigates whether certain organizational differences are correlated with differences in performance. Policy choices, such as the unbundling of the sector, the introduction of private ownership, the size of utilities, or the introduction of a separate regulatory authority, have been suggested as key steps in improving the overall performance of the electricity sector (see, for example, Bacon and Besant-Jones 2001). Because evidence of the benefits of power sector reform has not been overwhelming, policy responses to underperformance are being reevaluated.[2] Evidence of the impact of various reform strategies can help to inform the debate. The data collected for the analysis of performance in the MENA region contribute to this discussion.

In the present study, the availability of data drawn from a large number of utilities exhibiting different characteristics provides the opportunity to test for the effects of various reform strategies in a new way. If the average performance of all public utilities on various indicators is poorer than that of the average for all private utilities on the same indicators, then this supports the argument that privatization helps to improve performance. In making such comparisons we recognize that many factors contribute to performance on a given indicator, so that differences between public and private utilities are not due solely to ownership status. However, a significant difference in performance by ownership type would support the argument that ownership matters, whereas lack of significant difference would suggest that mode of ownership alone does outweigh all the other factors in determining performance on the indicator.

Part II focuses on detailed analysis of four countries that have taken very different approaches to the power sector: **Egypt, Jordan, Morocco,** and **Oman (chapters 6 to 9)**. Each country study provides an overview of the national power sector and an analysis of utility performance (comparing these with regional median values) to identify potential areas of improvement. The narrative and figures presented in these chapters focus on the year 2013.

Their characteristics and challenges of the case study countries are representative of the 14 MENA economies in this study, though each has a unique story to tell, whether related to its dependence on fossil fuel imports, its population and geographical size, or the initial and organizational structure of its electricity sector.

Several themes relevant to the region as a whole emerge from the case studies. First, all four countries have undertaken significant reforms of their electricity sectors over the past decades. Second, factors exogenous to the electricity sector have had an impact on utility performance. These include political instability; disruptions in primary-fuel supply; and spillovers from regional conflicts. Third, the four case-study countries have addressed in different ways the link between water and energy, a very salient matter in the MENA region. Fourth, some of the case studies deal with the introduction of renewable sources to the energy mix in a region where fossil fuels remain the dominant source of electricity.

As with the report as a whole, the case studies are limited by the availability of data. Yet they represent a good start toward the more consistent and developed analysis needed to meet the major challenges identified in this report. Key points raised in the case studies are presented, country by country, in *chapter 10*.

Notes

1. For example, the methodology used to compute utility QFDs in this chapter was largely inspired by Trimble and others (2016). Another example of the use of QFD is Eberhard and others (2008).

2. Vagliasindi and Besant-Jones (2013) show that unbundling can deliver performance improvements, but not for all indicators.

References

Bacon, R., and J. Besant-Jones. 2001. "Global Electric Power Reform, Privatization and Liberalization of the Electric Power Industry in Developing Countries." *Annual Review of Energy and Environment* 26 (1): 331–59. Also as: Energy and Mining Sector Board Discussion Paper 2, World Bank, Washington, DC.

Cambini, C., and D. Franzi. 2013. "Independent Regulatory Agencies and Rules Harmonization for the Electricity Sector and Renewables in the Mediterranean Region." *Energy Policy* 60 (September): 179–91.

Eberhard, A., V. Foster, C. Briceño-Garmendia, F. Ouedraogo, D. Camos, and M. Shkaratan. 2008. "Underpowered: The State of the Power Sector in Sub-Saharan Africa." Africa Infrastructure Country Diagnostic (AICD), summary of Background Paper No. 6, World Bank, Washington, DC.

Ianchovichina, E., A. Estache, R. Foucart, G. Garsous, and T. Yepes. 2012. "Job Creation through Infrastructure Investment in the Middle East and North Africa." Policy Research Working Paper No. 6164, World Bank, Washington, DC.

Trimble, C., M. Kojima, I. P. Arroyo, and F. Mohammadzadeh. 2016. "Financial Viability of Electricity Sectors in Sub-Saharan Africa: Quasi-Fiscal Deficits and Hidden Costs." Policy Research Working Paper 7788, World Bank, Washington, DC.

Vagliasindi, M., and J. Besant-Jones. 2013. *Power Market Structure: Revisiting Policy Options*. Directions in Development Series. Washington, DC: World Bank.

PART I

How Do MENA's Electricity Utilities Perform?

CHAPTER 1

Quasi-Fiscal Deficits in MENA's Power Sector

Estimating the power sector's quasi-fiscal deficit (QFD) provides a first attempt at quantifying the hidden costs originating from sector inefficiencies. When incurred by utilities over years, hidden costs can cause financial strain. This, in turn, can worsen an already delicate fiscal situation. A majority of economies in the Middle East and North Africa (MENA) region are at risk of financial strain.

The QFD, used in numerous analyses of various infrastructure sectors, including electricity, is a powerful tool for communicating with policy makers.[1] Four types of inefficiency contribute to the QFD:

- *Financial,* as measured by the size of the gap between the average tariff and the cost-recovery rate (*underpricing*)
- *Technical,* or the difference between actual transmission and distribution (T&D) losses and those of an ideal utility[2]
- *Commercial,* or the share of bills not collected (*collection losses*)
- *Labor,* as estimated by comparing the number of customers per utility employee against an efficiency benchmark[3] (*overstaffing*)

All four inefficiences can be expressed in absolute monetary terms or as a percentage of gross domestic product (GDP) or of a utility's revenue. Their calculation is illustrated in equation 1.1:

$$Q_e(T_c - T_e) + \frac{Q_e T_c (l_m - l_n)}{1 - l_m} + Q_e T_e (1 - R_{ct}) + \left(\frac{413 - \frac{NC}{NE}}{413} \right) CL \qquad (1.1)$$

$$\underbrace{}_{\substack{\text{Financial} \\ \text{inefficiency}}} \quad \underbrace{\phantom{\frac{Q_e T_c}{1}}}_{\substack{\text{Technical} \\ \text{inefficiency}}} \quad \underbrace{}_{\substack{\text{Commercial} \\ \text{inefficiency}}} \quad \underbrace{}_{\substack{\text{Labor} \\ \text{inefficiency}}}$$

Qe = end-user consumption (kilowatt-hours, kWh)

Tc = cost-recovery tariff ($/kWh)

Te = average end-user tariff ($/kWh)

$\ln$ = normative loss rate (%)

NC = number of customers

CL = cost of labor ($) per employee

$\mathrm{l}m$ = technical loss rate (%)

Rct = collection rate (%)

NE = number of employees

413 = benchmark number of customers per employee

This chapter seeks to estimate the hidden costs of the power sector in MENA. It represents the first time—to the best of our knowledge—that QFDs have been calculated in a context where utilities are not fully vertically integrated. We estimate QFDs at both the economy and utility level. In economies where there is only one vertically integrated utility (VIU), one might expect both estimates to be similar. But where the fuel used to generate electricity is subsidized, a significant gap between the two is created. Meanwhile, in economies with some degree of unbundling and more than one utility, the economy and utility-level QFDs differ as expected, and a number of methodological considerations and hypotheses need to be considered. In theory, the sum of the QFDs of an economy's utilities should equal the economy's own QFD. Although both QFDs can be expressed as a percentage of GDP, the utility-level QFD can also be expressed as a fraction of the utility's revenues. Of the 14 MENA economies considered in this report, all except the Arab Republic of Egypt, Jordan, Morocco, Oman, and the West Bank are treated as having just one VIU.

Several sources were used in an effort to acquire a maximum amount of data for the calculations of the economy- and utility-level QFDs. Most data come from the MENA Electricity Database, the World Development Indicators (WDIs), and reports from the Arab Union of Electricity. Appendix C provides the sources used for each indicator. This appendix also explains how several methodological challenges related to data availability were solved. Often, data for all 14 economies were not available in a single source, requiring further collection and verification. Particular challenges were faced, for instance, in gathering utilities' bill-collection rates—necessary to calculate the QFD's commercial inefficiency component—or estimating their labor costs in economies with several utilities.

Economy-Level Results

As can be seen in figure 1.1, half of the 14 MENA economies studied have a QFD above 4 percent of their GDP. Of particular concern is the fact that Lebanon, Djibouti, Bahrain, and Jordan have a QFD between 7.5 percent and 9 percent of GDP.

Table 1.1 lists QFDs for the 14 economies (expressed both in absolute terms and as a percentage of GDP) as well as the four individual components (as a percentage of GDP). Five economies have a QFD below 3 percent of GDP (West Bank, Morocco, Tunisia, Qatar, and Algeria); another four have a QFD

Figure 1.1 The Quasi-Fiscal Deficit as a Percentage of GDP, 14 MENA Economies, 2013

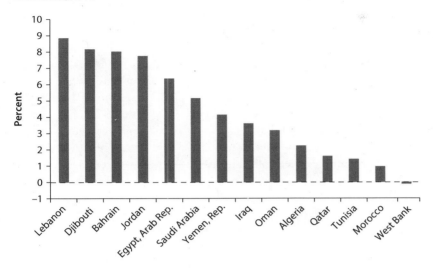

Source: World Bank calculations.
Note: GDP = gross domestic product; MENA = Middle East and North Africa.

Table 1.1 Quasi-Fiscal Deficit Calculations at the Economy Level, 2013 (except as noted)

Economy	Absolute QFD (US$ million)	QFD as share of GDP (%)	QFD components as share of GDP (%)			
			Underpricing	T&D losses	Collection losses	Overstaffing
Lebanon	3,826	8.9	8.20	0.41	0.21	0.03
Djibouti	101	8.2	0.98	1.08	5.24	0.88
Bahrain	2,640	8.0	7.86	0.02	0.02	0.13
Jordan	2,608	7.8	5.96	0.84	0.75	0.21
Egypt, Arab Rep.	18,219	6.4	5.61	0.42	0.06	0.28
Saudi Arabia	38,467	5.2	4.81	0.11	0.17	0.07
Yemen, Rep.	1,494	4.2	3.16	0.81	0.08	0.11
Iraq	7,888	3.6	2.44	0.83	0.13	0.21
Oman	2,496	3.2	2.70	0.22	0.18	0.10
Algeria	4,720	2.3	1.46	0.37	0.10	0.32
Qatar	3,224	1.6	1.47	0.02	0.10	0.01
Tunisia	655	1.4	0.34	0.39	0.54	0.15
Morocco	948	1.0	0.65	0.33	0.20	−0.21
West Bank	−13	−0.1	−0.84	0.30	0.30	0.13

Source: World Bank calculations.
Note: The year is 2013 for all except the following: 2012 for Lebanon, Iraq, Morocco, and the West Bank; and 2011 for Djibouti. This variation reflects data availability. GDP = gross domestic product; MENA = Middle East and North Africa; QFD = quasi-fiscal deficits; T&D = transmission and distribution.

between 3 percent and 6 percent of their GDP (Oman, Iraq, the Republic of Yemen, and Saudi Arabia), and five economies have a QFD between 6 percent and 9 percent of GDP (Egypt, Jordan, Bahrain, Djibouti, and Lebanon). In short, the QFD's share of GDP is relatively small in Maghreb economies, and large in some Mashreq and Gulf Cooperation Council (GCC) economies.

In absolute terms, the highest QFDs are to be found in Saudi Arabia ($38 billion), Egypt ($18 billion), and Iraq ($8 billion) and the lowest in the West Bank (with a negative QFD of $13 million), Djibouti ($101 million, despite having the second-highest QFD when expressed as a percentage of GDP), and Tunisia ($655 million). These values strongly correlate to the size of the economy and to the consumption levels of its population. As seen in table 1.1, we obtain negative values for overstaffing in Morocco and underpricing in the West Bank. This simply means that Morocco's ratio of customers to employees is better than the efficiency benchmark (413:1) used in this report, and that the West Bank's cost-recovery tariff is smaller than the average end-user tariff (based on the energy mix of Israel, given that the West Bank imports all its electricity from there).

Underpricing appears to be the main factor behind high QFD values: in 8 of the 14 economies, this component represents more than three-quarters of the QFD. In as many as 11 economies, it represents two-thirds. Underpricing does not, in itself, help disentangle the two common types of subsidies: that is, subsidies (a) of electricity and (b) of the fuels used to generate electricity. This is because the cost-recovery tariff used to estimate the economy-level QFD is based on levelized energy costs, computed as weighted averages of each economy's energy mix, to which a factor was added to account for T&D costs. The utility-level QFDs presented in table 1.2 and their comparison with economy-level QFDs allow us to better differentiate the two types of subsidies. The reason Djibouti and the West Bank are notable exceptions to the trend of underpricing as a driving force of the QFD is that both economies have high average end-user tariffs: $0.31 per kilowatt-hour (kWh) and $0.16 per kWh, respectively.

Technical inefficiencies (represented by T&D losses) are an important part of some economies' QFDs: they represent more than one-fifth of the total QFDs in Iraq, Morocco, Tunisia, the West Bank, and the Republic of Yemen. Commercial inefficiencies (represented by bill collection losses) represent as much as two-thirds of the QFD in Djibouti, more than one-third in Tunisia, and a substantial share in Morocco and the West Bank. Uncollected bills do not appear to be a key QFD component in the remaining 10 economies.

Finally, labor inefficiencies represent between 10 percent to 15 percent of the QFDs in Algeria, Tunisia, and Djibouti. Expressed as a percentage of GDP, they represent 1.0 percent in Djibouti and between 0.2 percent and 0.3 percent in Egypt and Jordan. Addressing this type of inefficiency may be a delicate act for governments, because it implies reducing the size of state-owned enterprises (SOEs). Providing public jobs—and subsidized basic services—has been part of the social contract in the region for the past several decades, in exchange for social stability.

How do economy QFD results for MENA compare with other regions? A recent study that computed QFDs for 17 Sub-Saharan African economies obtained values that ranged between −0.3 percent and 6.0 percent of GDP (Trimble and others 2016). Again, our values for MENA range from −0.1 percent in the West Bank to 8.9 percent in Lebanon (see table 1.1 for QFDs, including values for each of the four components, as a share of GDP). This indicates that MENA's utilities are more inefficient than Africa's: the median QFD in the 17 African economies is 0.8 percent of GDP, whereas it is close to 4.0 percent in the 14 MENA economies. Interestingly, although the MENA QFD appears to be driven mostly by financial inefficiency, in the case of Sub-Saharan Africa, technical and commercial inefficiencies play the largest role overall.

Another study (Ebinger 2006) of water and electricity sectors in 16 economies of Europe and Central Asia (ECA) found that tariffs set below cost-recovery rates were the biggest culprits behind the energy sector's hidden costs (the study did not consider overstaffing). QFDs as a share of GDP were as high as 14 percent in Tajikistan and 8 percent in Moldova—that is, comparable to the highest values in our sample of MENA economies.

Utility-Level Results

We computed QFDs for VIUs and distribution utilities (DUs) when sufficient data were available. This was done for all 14 economies but Qatar (because of insufficient information on that country's VIUs, the Qatar General Electricity and Water Corporation, KAHRAMAA). In total, QFDs were computed for 28 utilities, of which 11 are VIUs and 17 are DUs. A limitation of this exercise is that we were not able to compute QFDs for generation utilities (GUs) and transmission utilities (TUs). The reason is that, although we had data on the end-user tariffs set by VIUs and DUs (Arab Union of Electricity 2014), we did not have this data for GUs selling electricity (be it to a TU, a single buyer, or a VIU, depending on the market structure of the economy) or for TUs selling electricity. This gap prevented us from computing the financial inefficiency component at the utility level for GUs and TUs, which is a key component of the economy-level QFD in much of MENA.

The formula used to compute the QFD at the utility level is the same as that used at the economy level. However, two important differences merit clarification, because they drive much of the difference between the two types of QFD:

- Qe at the economy level is the end-user consumption (taken from the WDIs), whereas Qe at the utility level represents the amount of energy billed (taken from the MENA Electricity Database).
- Whereas Tc at the economy level is based on the energy mix and levelized costs of each generation technology (in addition to a T&D component), Tc at the utility level is based on the investment and operating costs of the utility annualized (taken from the MENA Electricity Database).

Shedding Light on Electricity Utilities in the Middle East and North Africa
http://dx.doi.org/10.1596/978-1-4648-1182-1

To compute the QFD of a utility, we used the year 2013 or, depending on data constraints, another year in the study period (2009–13). In some cases, to get around data gaps, we used observations across a few years. The methodology is described in appendix C, as well as the assumptions considered and sources of data. The results of our exercise—that is, utility-level QFDs, presented as a fraction of GDP and as a fraction of revenue of utilities—are presented in table 1.2.

The QFDs of VIUs tend to be higher than those of DUs. Of the 11 VIUs analyzed, 3 have a QFD above 4.0 percent of GDP (reaching as much as 8.4 percent in the case of Electricité de Djibouti), 6 are between 1.0 percent and 2.5 percent, and 2 are below 0.5 percent. These figures differ significantly from the 17 QFDs computed for DUs: 3 are at 0.8 percent of GDP or above (reaching as much as 2.3 percent in the case of the Jordan Electric Power Company), 6 are between 0.3 percent and 0.5 percent, and 8 are at 0.2 percent or below (and 6 of this last subset are in Egypt).

One way to adjust for the differences in types of utilities is to look at the QFD as a percentage of utility revenue, which provides revealing results. The Northern Electricity Distribution Company (NEDCO) in the West Bank is the DU with the lowest QFD as a proportion of its revenue (25 percent), whereas the Iraqi Ministry of Electricity's proportion is 1,267 percent (in other words, the monetary value of its inefficiencies is more than 12 times its revenue). Some other outliers on the upper end include Jerusalem District and Tubas District Electricity Companies (448 percent and 193 percent, respectively) and Electricité du Liban (372 percent). Of the 28 utilities analyzed, 13 have inefficiencies that are higher than their revenues. In other words, these utilities would double their revenue if they were to maximize their efficiency.

When one looks at the QFD components at the utility level, the results are—unsurprisingly—similar to those seen at the economy level: underpricing is by far the main driving force, except in Djibouti where collection losses play this role. For DUs, underpricing is the main driving force in Jordan, Morocco, and Oman. The West Bank is the exception, because T&D or collection losses are the driving force here. This is because the average end-user tariff in the West Bank is relatively high.

Table 1.3 compares the economy QFD with the utility QFD for economies with only one utility. The right-hand column discusses the observed differences and provides plausible explanations for them. In economies with one utility, we would expect to obtain similar values for both the economy-level and the utility-level QFD (with minor differences due to methodological considerations or different sources of data). However, in four of nine economies with one utility, we observe economy QFDs that are between 1.5 and 3.0 times higher than the utility QFD. Two main factors explain this: (a) the subsidies of fuel used for electricity generation (as in Saudi Arabia or Lebanon) artificially diminish the VIU's cost-recovery tariff, because the latter is based on artificially low operating

Table 1.2 Quasi-Fiscal Deficit Calculations at the Utility Level, Selected Utilities across MENA, 2013 (or most recent year with data, 2009–12)

Economy	Type	Utility	QFD as share of GDP (%)	QFD as share of revenues (%)	QFD components as share of GDP (%)			
					Underpricing	T&D losses	Collection losses	Overstaffing
Djibouti	Vertically integrated	Electricité de Djibouti	8.4	93	1.20	1.10	5.24	0.88
Lebanon	Vertically integrated	Électricité du Liban	5.7	372	5.27	0.26	0.11	0.03
Bahrain	Vertically integrated	Electricity and Water Authority	4.1	125	3.96	0.01	0.01	0.13
Tunisia	Vertically integrated	Société Tunisienne de l'Électricité et du Gaz	2.9	73	1.67	0.56	0.51	0.15
Jordan	Distribution	Jordan Electric Power Company	2.3	71	1.92	0.33	0.05	–0.01
Iraq	Vertically integrated	Ministry of Electricity	2.3	1,267	1.45	0.50	0.10	0.21
Algeria	Vertically integrated	Société Nationale de l'Électricité et du Gaz	1.9	129	1.16	0.33	0.10	0.32
Saudi Arabia	Vertically integrated	Saudi Electricity Company	1.7	131	1.40	0.04	0.17	0.07
Yemen, Rep.	Vertically integrated	Public Electricity Corporation	1.4	152	0.79	0.47	0.06	0.11
West Bank	Distribution	Jerusalem District Electricity Company	1.1	448	0.37	0.55	0.08	0.08
Morocco	Vertically integrated	Office National de l'Électricité et de l'Eau Potable	1.1	38	0.74	0.24	0.18	–0.08
Jordan	Distribution	Electricity Distribution Company	0.8	88	0.45	0.07	0.24	0.04
Morocco	Distribution	LYDEC	0.5	70	0.29	0.01	0.15	0.09
Oman	Distribution	Muscat Electricity Distribution Company	0.4	61	0.32	0.03	0.07	–0.01

table continues next page

17

Table 1.2 Quasi-Fiscal Deficit Calculations at the Utility Level, Selected Utilities across MENA, 2013 (or most recent year with data, 2009–12) *(continued)*

Economy	Type	Utility	QFD as share of GDP (%)	QFD as share of revenues (%)	QFD components as share of GDP (%)			
					Underpricing	T&D losses	Collection losses	Overstaffing
Oman	Distribution	Mazoon Electricity Distribution Company	0.4	64	0.31	0.03	0.05	0
Oman	Distribution	Majan Electricity Company	0.3	62	0.21	0.03	0.04	0.02
Egypt, Arab Rep.	Distribution	South Cairo Electricity Distribution Company	0.3	95	0.18	0.01	0.01	0.06
Oman	Vertically integrated	Rural Areas Electricity Company	0.3	121	0.20	0.01	0.01	−0.04
Egypt, Arab Rep.	Distribution	Canal Electricity Distribution Company	0.3	100	0.19	0	0	0.05
Egypt, Arab Rep.	Distribution	North Cairo Electricity Distribution Company	0.2	95	0.14	0.01	0	0.04
Oman	Vertically integrated	Dhofar Power Company	0.2	72	0.11	0.01	0.02	0
West Bank	Distribution	Tubas District Electricity Company	0.2	193	−0.03	0.01	0.11	0
West Bank	Distribution	Northern Electricity Distribution Company	0.2	25	0.03	0.05	0.06	0.02
Egypt, Arab Rep.	Distribution	Upper Egypt Electricity Distribution Company	0.1	125	0.08	0.02	0.01	0.03
Egypt, Arab Rep.	Distribution	Middle Egypt Electricity Distribution Company	0.1	103	0.08	0.01	0	0.03
Egypt, Arab Rep.	Distribution	Alexandria Electricity Distribution Company	0.1	115	0.07	0.01	0	0.04
Egypt, Arab Rep.	Distribution	North Delta Electricity Distribution Company	0.1	92	0.08	0.01	0.01	0.03
Egypt, Arab Rep.	Distribution	El-Behera Electricity Distribution Company	0.1	99	0.07	0.01	0	0.03

Source: World Bank calculations. Utilities selected based on data availability.

Note: GDP = gross domestic product; MENA = Middle East and North Africa; QFD = quasi-fiscal deficits; T&D = transmission and distribution.

Table 1.3 Comparison of Utility- and Economy-Level Quasi-Fiscal Deficits for Economies with One Utility, 2013 (or most recent year with data, 2009–12)

Economy	Utility	Economy QFD (% of GDP)	Utility QFD (% of GDP)	Factors explaining differences between the two QFDs
Algeria	Société Nationale de l'Electricité et du Gaz (SONELGAZ)	2.3	1.9	Only underpricing and T&D losses (economy QFD is higher by 0.3% and 0.04%, respectively). Specifically, T_c is different for both cases, given different sources and methodologies used (see appendix C for more details): it is 12.0¢/kWh for the economy, and 10.6¢/kWh for the utility. In addition, there is a minor difference in losses: 18.4% for the economy (World Development Indicators, or WDI) and 18.8% for the utility (MENA Electricity Database, or MED).
Bahrain	Electricity and Water Authority (EWA)	8.0	4.1	Underpricing, T&D losses, and collection losses (economy QFD is higher by 3.9%, 0.01%, and 0.01%, respectively). The main variable driving differences is Q_e, and T_c to a much lesser extent. Specifically, T_c is 11.3¢/kWh for the economy and 10.5¢/kWh for the utility; Q_e is 24.6 TWh for the economy (WDI), and 13.4 TWh for the utility (MED). The reason for this important difference is that EWA represents only 63% of installed capacity, because the other 37% (2,249 MW)[a] is the self-generation of an aluminum smelting company, Alba.
Djibouti	Electricité de Djibouti (EDD)	8.2	8.4	Underpricing and T&D losses, particularly underpricing, which is 0.2% higher for the utility-level QFD. This difference is driven by a slightly higher T_c for the utility (35.5¢/kWh) than for the economy (34.7¢/kWh). The small difference between the two values is due to methodology. Both QFDs correspond to the year 2011 because there were insufficient data for 2013.
Iraq	Ministry of Electricity (MOE)	2.5	2.3	Underpricing (0.18% higher in the economy) and to a smaller extent T&D losses. The main factor driving these differences is T_c, which is 11.9¢/kWh for the economy and 9.9¢/kWh for the utility. Both QFDs correspond to the year 2012 because of insufficient data for 2013.
Lebanon	Electricité du Liban (EdL)	8.9	5.7	Differences observed in all but overstaffing components, but mainly in underpricing (economy QFD is 2.93% higher). This is primarily driven by T_c and Q_e differentials. T_c is 29.0¢/kWh for the economy and 34.8¢/kWh for the utility. Q_e is 7.2 TWh for the utility (2012) and 13.8 TWh for the economy (2012). The difference in Q_e can be attributed to the fact that about 35% of energy consumption in Lebanon is self-generation for own consumption.[b] Both QFDs correspond to the year 2012 because of insufficient data for 2013.
Qatar	Qatar General Electricity and Water Corporation (KAHRAMAA)	1.6	—	n.a.

table continues next page

19

Table 1.3 Comparison of Utility- and Economy-Level Quasi-Fiscal Deficits for Economies with One Utility, 2013 (or most recent year with data, 2009–12) *(continued)*

Economy	Utility	Economy QFD (% of GDP)	Utility QFD (% of GDP)	Factors explaining differences between the two QFDs
Saudi Arabia	Saudi Electricity Company (SEC)	5.2	1.7	Underpricing (economy-level QFD higher by 3.41%) and to a lesser extent T&D losses (economy QFD higher by 0.07%). The difference is driven mainly by different Tc values: 14.9¢/kWh for the economy and 5.4¢/kWh for the utility. This significant differential can be attributed to the importance of fuel subsidies, which are included in the economy Tc but not in the utility Tc, because these subsidies would not be reflected in the OPEX of the utility.
				In 2013, the fuel subsidy for electricity generation given to SEC was US$40 billion.[c] If we were to add this value to the OPEX of SEC, we would then see an increase in its Tc (5.4¢/kWh to 21.0¢/kWh), and consequently the utility QFD would also increase (from 1.7% to 7.2%) to a value much closer to the economy QFD. The remaining differences between the economy and utility QFDs can be attributed to methodology.
Tunisia	Société Tunisienne de l'Electricité et du Gaz (STEG)	1.4	2.9	Underpricing (1.3% higher in utility), and to some lesser extent T&D losses (0.2% higher in utility). Tc is the main driver in differences because it is 11.2¢/kWh for the economy and 15.5¢/kWh for the utility.
Yemen, Rep.	Public Electricity Corporation (PEC)	4.2	1.4	Differences observed in all but overstaffing components, and most in underpricing (2.37% higher in the economy), and to a lesser extent T&D losses (0.34% higher in the economy). Tc is the main driver: 21.1¢/kWh for the economy[d] and 8.8¢/kWh for the utility. This significant differential can be attributed to the importance of fuel subsidies, which are included in the economy Tc but not in the utility Tc, as these subsidies would not be reflected in the OPEX of the utility.
				Fuel subsidies for electricity generation given to PEC amounted to about $1.1 billion.[e] If we were to add this value to the OPEX of PEC, we would then see an increase in its Tc (8.8¢/kWh to 30.6¢/kWh), and consequently the utility QFD would also increase (from 1.4% to 5.6%). The remaining differences between the economy and utility QFD can be attributed to methodology.

Source: World Bank calculations. Specific sources and additional information in notes.

Note: GDP = gross domestic product; kWh = kilowatt-hour; MENA = Middle East and North Africa; MW = megawatts; OPEX = operating expenses; Qe = end-user consumption (kWh); QFD = quasi-fiscal deficit; T&D = transmission and distribution; Tc = cost-recovery tariff (US¢/kWh); TWh = terawatt-hours; — = negligible (insufficient data available); n.a. = not applicable.

a. Alba's website: http://www.albasmelter.com/About%20Alba/Factsfigures/Pages/default.aspx.

b. World Bank 2009, 18.

c. Jeddah Chamber of Commerce 2015, 11.

d. Of the 68 percent of electricity that is fuel (WDI), we assume that half (34 percent of total) is based on diesel self-generation, and that the remaining half (34 percent of total) is produced by PEC, using in equal amounts heavy fuel oil (HFO) and diesel, as per the Arab Union of Electricity (AUE) Manual of Power Stations.

e. Fattouh and El-Katiri 2012: 30. Although this figure corresponds to 2008, the cost of fuel purchases of PEC remained stable between 2009 and 2012 according to PEC reported figures.

expenses; and (b) even in economies with a vertically integrated electricity market, self-generation can be widespread among residential consumers (as in Lebanon or the Republic of Yemen) or even industrial consumers (as with an aluminum smelting company in Bahrain).

Table 1.4 compares the economy-level QFD with the utility-level QFD for economies with multiple utilities. In such economies, we expected to see higher values for economy QFDs than for the sum of utilities' QFDs because we did not have sufficient data to compute the QFDs of all the utilities present. This assumption holds true in the case of Egypt, Jordan, and Oman. Interestingly, in the West Bank and in Morocco we see the opposite happening. In the West Bank, this could be because all the electricity consumed is imported from Israel, and the selling price may be delinked from the levelized cost of energy in Israel. In Morocco, this could be attributed to understaffing: Morocco's Office National de l'Electricité et de l'Eau Potable (ONEE) has one employee per 557 customers, higher than the benchmark of 413. In addition to being a VIU, ONEE is also the single buyer of electricity in Morocco and sells mainly to the 11 DUs.

Table 1.4 Comparison of Economy- and Utility-Level Quasi-Fiscal Deficits for Economies with Multiple Utilities, 2013 (or most recent year with data, 2009–12)

Economy	Economy QFD (% GDP)	Utility QFD (% GDP)	Factors explaining differences between the two types of QFDs
Egypt, Arab Rep.	6.4	1.4	The QFD of utilities corresponds only to the nine Egyptian DUs, because data were not sufficient to compute it for the TUs and GUs. Tc for the economy is 12.6¢/kWh, and for the DUs it oscillates between 3.8¢/kWh and 4.5¢/kWh. This presumably indicates that the electricity that DUs buy is subsidized.
Jordan	7.8	2.8	The QFD of utilities corresponds to only two Jordanian DUs, because data were not sufficient to compute it for the one remaining DU, the six GUs, and the one TU. Tc for the economy is 19.8¢/kWh, and for the two DUs it is 12.4¢/kWh and 10.9¢/kWh. This presumably indicates that the electricity that the DUs buy is subsidized.
Morocco	1.0	1.5	The utility QFD corresponds to the vertically integrated utility ONEE because data for the 11 DUs were insufficient. The Tc for the vertically integrated utility ONEE, which is the single buyer of electricity and generates 42% of the total electricity supplied, is 4.4¢/kWh, although the economy Tc is 3.6¢/kWh. As a single buyer, ONEE sells about 44% of its electricity to the DUs, at an average price of 10.5¢/kWh[a], and the rest is sold directly to consumers (average tariff is calculated at 11.3¢/kWh). Both QFDs correspond to 2012 because data were not sufficient for 2013.

table continues next page

Shedding Light on Electricity Utilities in the Middle East and North Africa
http://dx.doi.org/10.1596/978-1-4648-1182-1

Table 1.4 Comparison of Economy- and Utility-Level Quasi-Fiscal Deficits for Economies with Multiple Utilities, 2013 (or most recent year with data, 2009–12) *(continued)*

Economy	Economy QFD (% GDP)	Utility QFD (% GDP)	Factors explaining differences between the two types of QFDs
Oman	3.2	1.3	The QFD of utilities corresponds to only one VIU and three DUs, because data were not sufficient to compute it for the remaining one DU, one TU, one VIU, and 12 GUs. *Tc* for the economy is 1.6¢/kWh, and *Tc* for utilities oscillates between 5.4¢/kWh and 6.7¢/kWh, except for the rural VIU for which the *Tc* goes up to 27.0¢/kWh. The latter is presumably due to the usage of diesel generators to produce electricity. Here again, because the *Tc* for the economy tends to be above the one for DUs, presumably the electricity that the DUs buy is subsidized.
West Bank	−0.1	1.4	It is striking that the economy-level QFD is slightly negative. This result is driven by a negative underpricing component (−0.8% of GDP), because *Te* is 16.4¢/kWh whereas *Tc* is 11.4¢/kWh. This result is surprising, and may reflect the fact that the cost at which DUs in the West Bank are buying electricity from Israel is delinked from costs. The rest of the QFD economy components oscillate between 0.1% and 0.3% of GDP.
			The QFDs of utilities corresponds to three DUs, which are all the utilities in the West Bank in this study (most generation comes from Israel). In theory, given that our utility QFD covers all utilities in the West Bank, both types of QFD should be equal. The reason this is not so is because the *Tc* of utilities is considerably higher and oscillates between 12.7¢/kWh and 19.3¢/kWh.
			Both QFDs correspond to 2012 because data were not sufficient for 2013.

Source: World Bank calculations.
Note: DU = distribution utility; GDP = gross domestic product; GU = generation utility; kWh = kilowatt-hours; ONEE = Office National de l'Electricité et de l'Eau Potable; QFD = quasi-fiscal deficit; *Tc* = cost-recovery tariff (¢/kWh); *Te* = average end-user tariff (¢/kWh); TU = transmission utility; VIU = vertically integrated utility.
a. 0.88 Moroccan dirham per kilowatt-hour average.

What Can Be Done about Underpricing in MENA Economies?

Underpricing is by far the biggest factor behind the high QFD values observed in the power sector of the MENA region. This is not surprising when looking at figure 1.2, because the average end-user tariff appears to be below the cost-recovery level in all economies but the West Bank. In some cases, this differential is particularly acute—Lebanon and the Republic of Yemen being primary examples. Other economies with relatively high average electricity tariffs are Tunisia, Morocco, and Djibouti.

MENA average end-user tariffs are low when compared to the rest of the world. Table 1.5 lists tariffs for MENA economies and basic statistics for economies both in and outside MENA (the latter based on a sample of more than 60 countries). The average and median of non-MENA economies are more than twice and four times the respective MENA values.

Figure 1.2 Comparison of Average End-User and Cost-Recovery Tariffs in MENA, 2013 (or most recent year with data, 2009–12)

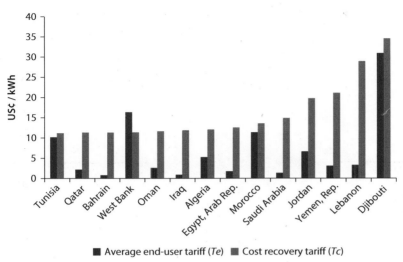

■ Average end-user tariff (*Te*)   ■ Cost recovery tariff (*Tc*)

Source: World Bank calculations.
Note: MENA = Middle East and North Africa; kWh = kilowatt-hour.

Table 1.5 Average Electricity Tariffs for MENA Economies and Comparison with Non-MENA Economies

Economy	¢/kWh	Economy	¢/kWh
Algeria	5.25	Morocco	11.36
Bahrain	0.8	Oman	2.6
Djibouti	31.0	Qatar	2.2
Egypt, Arab Rep.	1.78	Saudi Arabia	1.33
Iraq	0.9	Tunisia	10.19
Jordan	6.63	West Bank	16.38
Lebanon	3.29	Yemen, Rep.	3.1
MENA basic statistics		*Non-MENA basic statistics*	
Average	6.92	Average	15.62
Quartile 1	1.89	Quartile 1	8.5
Median	3.2	Median	13.55
Quartile 3	9.3	Quartile 3	18.98
Number of observations	14	Number of observations	61

Source: World Bank calculations.
Note: MENA tariffs come from the Arab Union of Electricity 2014; they represent the average domestic tariff of 250 kWh per month (in U.S. cents per kilowatt-hour). Non-MENA tariffs are obtained from Readiness for Investment in Sustainable Energy (RISE) and are mainly from sub-Saharan Africa, Asia, and Latin America. See Banerjee and others (2016). MENA = Middle East and North Africa; ¢/kWh = U.S. cents per kilowatt hour.

Tariff reforms could help improve the political viability of efforts to increase cost-recovery rates. Improved cost recovery, subsidy cuts, and better targeting demand a detailed look at the current design of electricity tariffs. Table 1.6 hints at the fact that, in many of the economies, there is at least some scope to improve the tariff design. Cross-subsidies do not systematically favor the poorest, even if

Table 1.6 Drivers of Electricity Tariff Design in MENA Economies

Economy	Peak/off-peak rates	Use	Seasonality	Surcharge for heavy nonresidential users	Lowest tariff	Cross-subsidies
Algeria	Yes	Yes	Yes	Yes	—	—
Bahrain	No	Yes	No	No	Residential	Yes
Egypt, Arab Rep.	No	Yes	No	No	Agriculture	Yes
Iraq	No	Yes	No	No	—	No
Saudi Arabia	Yes	Yes	Yes	No	Agriculture	Yes
Kuwait	No	No	No	No	—	No
Jordan	Yes (afternoon)	Yes	No	Yes	Industry	Yes
Lebanon	No	Yes	No	No	Residential and commercial	Yes
Libya	No	(Only in residential)	No	No	Residential	Yes
Morocco	Yes	Yes	Yes	No	Agriculture	Yes
Oman	Yes	Yes	Yes	No	—	No
Qatar	No	Yes	No	No	Agriculture and residential	No
Syria	Yes (evening)	Yes	No	No	Agriculture and public administration	Yes
Tunisia	Yes	Yes	Yes	Yes	Residential	Yes
United Arab Emirates	No	Yes (mostly)	No	No	Residential and commercial	Yes
West Bank	No	No	No	No	Industry	Yes
Yemen, Republic	No	Yes	No	No	Residential	Yes

Source: Arab Union of Electricity 2014 and utilities' websites.
Note: MENA = Middle East and North Africa. — = not available.

most economies seem to protect farmers and to some extent residential users. If the priority is to cut direct subsidies, then progressive cross-subsidies across user types might offer a way to promote joint efficiency, equity, financial viability, and, when needed, fiscal sustainability. Estimating the scope for improving tariff design would demand a much more thorough analysis than table 1.6 provides. But even though this book largely focuses on the supply side of the electricity business, there are other relevant dimensions to consider. In particular, there is a need to pay equivalent attention to the demand side, including to consumers' ability and willingness to pay. There is a good case for assessing the extent to which pricing options could do better to improve incentives on both sides of the market. International experience suggests that tariff reforms are likely to be part of a politically viable financing solution, in addition to management improvements needed to reduce the cost inefficiencies documented in the following chapters.

We have seen in the previous section that fuel subsidies for electricity generation are also an important part of the underpricing challenge in some MENA economies. For example, the QFD of Saudi Arabia increases by more than 5 percent if we account for fuel subsidies for electricity generation— equivalent to US$40 billion. The Republic of Yemen's increases by more than 4 percent—equivalent to US$1.1 billion. This represents by far the highest source of inefficiencies in these economies' electricity sectors. We did not have sufficient data to study the impact of subsidies in more detail, leaving an important topic for future analysis.

Conclusion

The median QFD value in the 14 MENA economies analyzed here is about 4 percent of GDP. This represents more than the average investment needed in the region's electricity sector, estimated at about 3 percent of GDP. In other words, the sector's investment gap could be filled simply by removing a fraction of the current level of inefficiency. Indeed, there is heterogeneity across the region, as QFD estimates vary from 8 percent to 9 percent of GDP in Lebanon, Djibouti, or Bahrain to less than 1.5 percent in Tunisia, Morocco, and the West Bank.

Underpricing appears to be the main driver of QFD for most economies in the region. This is due to the significant presence of subsidies for electricity and for the fuel used to generate electricity. Also, the cost-recovery tariff in many countries is high due to the significant presence of fuel in the energy mix. The other components of the QFD are T&D losses, collection losses, and overstaffing. These should not be forgotten, their aggregate median value for the 14 MENA economies is 0.8 percent of GDP, but it goes as high as 7.2 for Djibouti for example. Different priorities will need to be identified in different countries to reduce the QFD, incorporating the political economy in certain measures, be it tariff reforms or managing the sector's labor force.

To the best of our knowledge, this chapter represents the first effort to compute and compare QFDs at both the economy and at the utility levels. One advantage of this dual exercise is that the utility-level QFD is useful to utility managers, particularly when there are multiple electricity utilities in a given economy. Another advantage is that even without explicitly computing fuel subsidies for electricity production, we can get a sense of their size by comparing the results of the two types of QFD.

Finally, this chapter suffered from several data constraints. The most obvious is that we did not have data on the prices at which GUs were selling electricity, which prevented us from computing QFDs for these utilities. Another limitation was the quality of the data collected, particularly for variables such as the bill-collection rate or the cost of labor. Obtaining disaggregated data on the number of employees and the revenues in multiservice utilities also proved to be challenging, for example, in the case of Electricity and Water Authority in Bahrain, Société Tunisienne de l'Electricité et du Gaz in Tunisia, ONEE in Morocco, and Société Nationale de l'Electricité et du Gaz in Algeria.

Notes

1. For example, the methodology used for the QFD in this chapter has been greatly inspired by Trimble and others (2016). Another example of the use of QFD is Eberhard and others (2008).

2. T&D is fixed at 5 in this report because the best-performing utilities in our sample are slightly above this value, which we consider to correspond to an "ideal utility."

3. This inefficiency is estimated at 413 for developing countries, following Trimble and others (2016).

References

Arab Union of Electricity. 2014. *Electricity Tariff in the Arab Countries*. Statistical bulletin. Amman: Arab Union of Electricity.

Banerjee, S. G., A. Moreno, J. Sinton, T. Primiani, J. Seong. 2016. "Regulatory Indicators for Sustainable Energy: A Global Scorecard for Policy Makers." Working paper 112828, World Bank, Washington, DC.

Eberhard, A., V. Foster, C. Briceño-Garmendia, F. Ouedraogo, D. Camos, and M. Shkaratan. 2008. "Underpowered: The State of the Power Sector in Sub-Saharan Africa." Africa Infrastructure Country Diagnostic, summary of background paper 6, World Bank, Washington, DC.

Ebinger, J. O. 2006. "Measuring Financial Performance in Infrastructure: An Application to Europe and Central Asia." Policy Research Working Paper 3992, World Bank, Washington, DC.

Fattouh, B., and L. El-Katiri. 2012. "Energy Subsidies in the Arab World." Arab Human Development Report Research Paper Series, United Nations Development Programme, New York.

Jeddah Chamber of Commerce. 2015. *Sectorial Report on Saudi Arabia—Electricity July 2015*. Jeddah: Jeddah Economic Gateway. http://www.jeg.org.sa/data/modules/contents /uploads/infopdf/2832.pdf.

Prasad, T. V. S. N., M. Shkaratan, A. K. Izaguirre, J. Helleranta, S. Rahman, and S. Bergman. 2009. *Monitoring Performance of Electric Utilities: Indicators Benchmarking in Sub-Saharan Africa*. Washington, DC: World Bank. https://www.esmap .org/sites/esmap.org/files/P099234_AFR_Monitoring%20Performance%20of%20 Electric%20Utilities_Tallapragada_0.pdf.

Trimble, C., M. Kojima, I. P. Arroyo, and F. Mohammadzadeh. 2016. "Financial Viability of Electricity Sectors in Sub-Saharan Africa: Quasi-Fiscal Deficits and Hidden Costs." Policy Research Working Paper 7788, World Bank, Washington, DC.

World Bank. 2009. *Energy Efficiency Study in Lebanon—Final Report*. Washington, DC: World Bank. http://climatechange.moe.gov.lb/viewfile.aspx?id=205.

CHAPTER 2

Comparing the Region's Performance with the Rest of the World

This chapter provides a snapshot of key performance indicators for those power utilities in the Middle East and North Africa (MENA) region for which both regional and international comparisons are possible. The MENA data are taken from the MENA Electricity Database (MED), which itself is based on question-naires administered to 67 power utilities in the region. Data are for 2013 except in a few cases where another year between 2009 and 2012 was chosen because of data constraints. The MED covers 12 VIUs, 23 generation utilities (GUs), 29 distribution utilities (DUs), and 3 transmission utilities (TUs) (names and corre-sponding abbreviations for the utilities considered are in appendix B).

Ideally, each indicator would be compared across an identical set of utilities within MENA and also against a single set of utilities in countries from other regions. However, this ideal is not attainable at present. The database has varying coverage of the 36 indicators included in the survey for two reasons. First, certain indicators are relevant only to certain types of utilities: for example, generators do not experience distribution losses and hence do not collect such data. Second, many utilities did not provide data on certain indicators, even though these are relevant to utility performance: for example, only 49 of the 67 utilities surveyed reported data on their return on assets (ROA).

For data on non-MENA economies the challenge is greater: surveys cover only a subset of the MENA indicators, and this coverage differs across regions. Added to this is the problem of missing data. As a result, the sample of countries and utilities available[1] for comparison varies by indicator. To some extent the choice of indicator for global comparisons was dictated by data availability.

MENA and non-MENA performance was compared using the first (Q1), second (Q2), and third (Q3) quartiles of the data. The second quartile (median) denotes the level of the indicator at which one half of the observations were smaller and one half were larger. The median is preferred to the mean value because the latter is too sensitive to the dispersion of performance levels and the possible existence of extreme outliers in the data. Q1 separates the smallest 25 percent from the other observations, and Q3 separates the largest 25 percent from the other observations. This reduces the impact of extreme observations, which are given equal weight, relative to other observations smaller than Q1 or larger than Q3. The use of Q1 and Q3, in addition to Q2, which is close to the mean, allows a more complete comparison. For instance, the medians for two groups may be close, but the upper quartile may be considerably larger for one region, indicating that the best performers are not comparable. For some indicators, a high value denotes good performance and a low value poor performance (for example, ROA), whereas for other indicators, a high value denotes poor performance (for example, distribution losses).

Extensive statistical testing described in chapter 5 indicates that a substantial number of performance indicators are related to a country's income level (at higher incomes, performance is better). As a result, comparisons between regions may also be affected by differences in regional income levels. For example, income levels in Sub-Saharan Africa are generally well below those of the MENA region, whereas those in Latin America and the Caribbean (LAC) are nearer to the MENA levels. Hence the median values of performance indicators for Africa are likely to be lower than median levels for MENA. This comparison suggests that median MENA performance might be close to the Q3 performance of non-MENA utilities, allowing for the effect of income levels (for indicators where high values indicate high performance). This effect is particularly the case for vertically integrated utilities (VIUs) because of the predominance of utilities from low-income economies in Sub-Saharan Africa.

Comparisons—based on median, Q1, and Q3 values—are made for technical, financial, and commercial indicators to highlight possible differences in performance across regions. Further, MENA utilities have been compared with one another to highlight strong and weak performers (for the indicator in question).

For some indicators (for example, those related to operating expenses [OPEX] or total costs), performance cannot be expected to be the same for different types of utilities. For example, a VIU bears generation and transmission costs as well as the distribution costs borne by a DU supplying the same number of customers. However, data for VIUs cannot be disaggregated into various functions, so no comparison can be made between DUs and VIUs for such indicators. For other indicators, such as the ROA, performance depends entirely on efficiency, and so all utility types can be directly compared. The comparison between MENA and non-MENA utilities is split between DUs and vertically integrated utilities only when the indicators have different definitions, depending on the type of utility. In the case of GUs and TUs, there were too few observations to make adequate comparisons both within MENA and across global regions.

Summary of Results and Overall Assessment

Table 2.1 summarizes the relative performance of MENA versus non-MENA economies by comparing the median values of indicators across all utilities for which data were available. If an indicator is such that different utility structures can be expected to have similar performance levels, then the median values are used across all the utility types possible. If an indicator is such that performance varies by type, then comparisons are contained accordingly.

The main findings can be summarized as follows:

- MENA economies tend to perform better than non-MENA ones for about half of the indicators selected.
- The financial health of the MENA utilities is questionable. For example, the MENA value for accounts receivable over sales is almost three times that of non-MENA economies, and the ratio of current assets to current liabilities is lower than the non-MENA median—and lower than 100 percent. This is reinforced by a very high debt-to-equity ratio (almost four times the non-MENA median), leaving utilities highly exposed to external shocks. Although the ROA is above the non-MENA level, it is still low, suggesting that improvements in performance are required.

Table 2.1 Comparing the Median Performance of Selected MENA and Non-MENA Utilities, 2013 (or most recent year with data, 2009–12)

Indicator	Utility type	MENA median	Non-MENA median	MENA is superior?
OPEX/connection ($)	Distribution	346	129	No
OPEX/connection ($)	Vertically integrated	1,237	594	No
OPEX/kWh ($)	Distribution	0.10	0.14	Yes
OPEX/kWh ($)	Vertically integrated	0.07	0.18	Yes
Residential connections/employee	Distribution	252	367	No
Residential connections/employee	Vertically integrated	90	157	No
Distribution losses (%)	All	11	12	Yes
Energy sold/connection (kWh)	All	4,223	3,405	Yes
Total billing/connection ($)	All	299	292	Yes
Collection rate (%)	All	92	94	No
Sales/OPEX (%)	Distribution	93	98	No
Sales/OPEX (%)	Vertically integrated	92	87	Yes
Sales/total costs (%)	Distribution	88	67	Yes
Sales/total costs (%)	Vertically integrated	56	67	No
Accounts receivable/sales (days)	All	148	52	No
Debt/equity (%)	All	357	91	No
Current assets/current liabilities (%)	All	84	88	No
Return on assets (%)	All	3	1	Yes
Return on equity (%)	All	6	0	Yes

Source: World Bank calculations.
Note: kWh = kilowatt-hours; MENA = Middle East and North Africa; OPEX = operating expenses.

Shedding Light on Electricity Utilities in the Middle East and North Africa
http://dx.doi.org/10.1596/978-1-4648-1182-1

• One other area where there is a large difference between MENA and non-MENA economies is that of connections per employee. The low ratio in MENA suggests that hiring practices in MENA may need to be reviewed in some cases.

Detailed Comparisons for Selected Indicators

A more complete comparison between MENA and non-MENA economies is carried out in tables 2.2 to 2.15 by using indicator values at the Q1, Q2, and Q3 levels, both for all utilities and for VIUs, DUs, TUs, and GUs separately, where appropriate. The values of the indicators for all comparable utilities within MENA (for which data were available) are plotted and compared against the MENA and non-MENA median values.

Technical and Operational Performance Indicators

OPEX per connection ($). OPEX consists mainly of fuel costs, labor costs, maintenance and repair costs, and energy purchases. The relative proportions of these categories can be expected to vary among economies depending on relative prices. In particular, wage rates may vary to a large degree, while unit fuel costs may be similar (though, in fact, these costs can vary considerably in MENA given the importance of subsidy schemes in some economies). For instance, an inefficient utility might be overstaffed because of hiring practices or have excessive fuel bills because of poor dispatch decisions. Expenditure on maintenance may be inadequate because of a desire to cut costs in the short run. If labor costs dominate, then we expect that a more efficient utility would have lower OPEX per connection, other factors being held constant. Table 2.2 presents the data on OPEX per connection. This indicator does not apply to GUs, so those are excluded. Also, VIUs and DUs are separated because, as previously noted, a VIU incurs generation and transmission OPEX in addition to distribution costs.

Table 2.2 OPEX per Connection for MENA and Non-MENA Utilities, 2013 (or most recent year with data, 2009–12)

Utility type	Region	Number of utilities	Quartile 1: best performers ($)	Median ($)	Quartile 3: worst performers ($)
All	MENA	36	215	411	864
	Non-MENA	70	75	172	416
Distribution	MENA	25	157	346	547
	Non-MENA	48	67	129	197
Vertically integrated	MENA	11	665	1,237	1,547
	Non-MENA	22	243	594	852

Source: World Bank calculations.
Note: MENA = Middle East and North Africa; OPEX = operating expenses. The table indicates that OPEX per connection is less than $346 in 50 percent of distribution utilities inside MENA and less than $129 in 50 percent outside MENA. Notably, OPEX per connection is more than $547 in 25 percent of distribution utilities in MENA.

The medians are substantially higher for DUs and VIUs inside MENA than outside the region. The best-performing groups in MENA (Q1) and the worst-performing groups (Q3) are also much more pronounced than their non-MENA equivalents. These results suggest that the components of OPEX are substantially larger than their equivalents in non-MENA economies. In the case of VIUs, this could be explained by different fuel mixes used to generate electricity: OPEX for fuel-based generation are higher than for hydro-based generation, and fuel is important to many MENA economies. However, this cannot apply to the differences seen in DUs, which could also be due to higher wage bills in MENA.

Figure 2.1a plots the values of OPEX per connection for DUs in MENA. The values for the Muscat Electricity Distribution Company (MEDC) and the Mazoon Electricity Company (MZEC), both in Oman, stand out as being very high, given the amount of variation across the other utilities in MENA, whereas the values for the Egyptian utilities are all low. The latter could indicate an insufficient level of maintenance. Further, consumption levels explain part of these differences: for example, electricity consumption per capita in Oman is almost four times higher than that in the Arab Republic of Egypt. Differences in wage levels could also be a factor. It would be necessary to examine the reasons for these findings before concluding that the Oman utilities are unusually inefficient or the Egyptian utilities are very efficient.[2]

Figure 2.1b plots the values for the VIUs in MENA. Oman's Rural Areas Electricity Company (RAECO) is an outlier; its high value can be explained by the fact that it covers only rural areas, whereas the other VIUs listed also cover urban areas and benefit from economies of scale. Here again, differences could be explained by (a) differences in fuel mix (for example, fuel is critical to Djibouti's economy, and the value for that country's VIUs is above the median), (b) differences in salaries (for example, none of the VIUs in the Gulf Cooperation Council [GCC] economies has a value below the median) and levels of consumption (which is very low in the Republic of Yemen, for example).

OPEX per kWh sold ($). Low values of this indicator generally suggest relatively high levels of efficiency because the utility can supply a given amount of electricity at a relatively low operating cost. Where maintenance and repairs are obviously suboptimal, a low value of this indicator reflects inefficiency. In the case of VIUs, a low value can mask a utility's energy mix, specifically whether it is more or less based on fossil fuel. Table 2.3 presents the calculations of OPEX per kWh sold, divided by quartile for utilities both in and beyond MENA. For DUs, the median value is lower inside MENA than outside, and this trend is even more defined among the worst performers (Q3). These results suggest that utilities inside MENA have been better able to hold down costs per kWh than utilities outside the region. Meanwhile, as has been noted, OPEX per connection is higher inside MENA, where more kWh are supplied per connection than in other regions. Although the same trends are observed for VIUs, these

Figure 2.1 OPEX per Connections for Distribution and Vertically Integrated Utilities in MENA ($), 2013 (or most recent year with data, 2009–12)

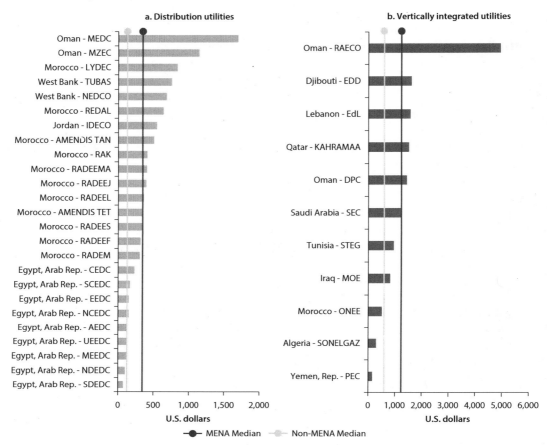

Source: MENA Electricity Database and World Bank calculations.
Note: MENA = Middle East and North Africa; OPEX = operating expenses.

Table 2.3 OPEX per kWh Sold for MENA and Non-MENA Utilities, 2013 (or most recent year with data, 2009–12)

Utility type	Region	Number of utilities	Quartile 1: best performers ($)	Median ($)	Quartile 3: worst performers ($)
All	MENA	36	0.04	0.10	0.13
	Non-MENA	28	0.08	0.14	0.24
Distribution	MENA	26	0.04	0.10	0.13
	Non-MENA	19	0.06	0.14	0.23
Vertically integrated	MENA	10	0.05	0.07	0.18
	Non-MENA	6	0.09	0.18	0.25

Source: World Bank calculations.
Note: kWh = kilowatt-hours; MENA = Middle East and North Africa; OPEX = operating expenses.

results should be treated with caution, keeping in mind that the sample of non-MENA VIUs is small.

Figure 2.2a shows the values of OPEX per kWh for DUs in MENA. The values for Lyonnaise des Eaux de Casablanca (LYDEC) in Morocco and the Jerusalem District Electricity Company (JDECO) in the West Bank are notably high. As is the case for OPEX per connection, Egypt's utilities have much lower values than those in other economies, a finding that requires further research to understand.

Figure 2.2b plots the values of OPEX per kWh for VIUs in MENA. The value for Electricité de Djibouti (EDD) in Djibouti is a clear outlier, and Electricité du Liban (EdL) in Lebanon and RAECO in Oman are also well above the median values for both MENA and non-MENA economies. The values for EDD and EdL may be explained by exceptionally high fuel costs and various types of inefficiencies. RAECO's value could be explained because it covers only rural areas.

Figure 2.2 OPEX per Kilowatt Hour Sold ($), MENA, 2013 (or most recent year with data, 2009–12)

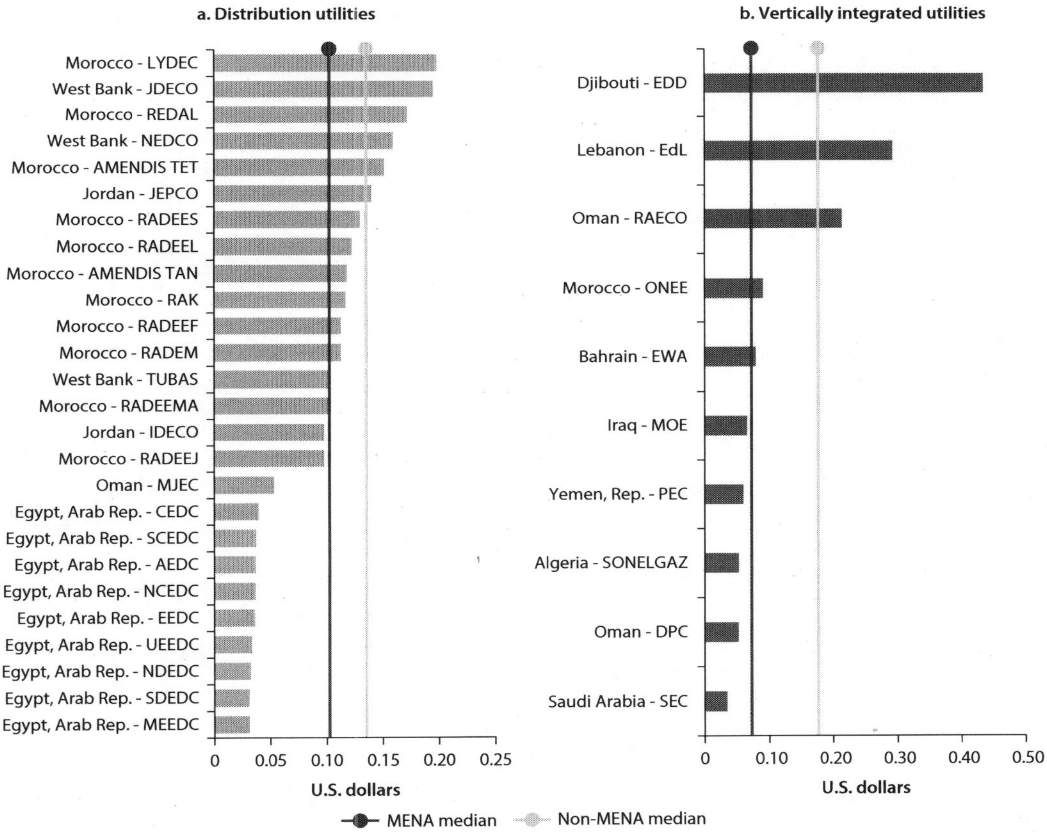

Source: MENA Electricity Database and World Bank calculations.
Note: MENA = Middle East and North Africa; OPEX = operating expenses.

Residential connections per full-time equivalent employee (FTE). Relatively efficient utilities are expected to sustain a greater number of connections per employee. A VIU would be expected to have a lower value than a DU with the same number of connections because supplying the additional generation and transmission requires extra labor. Table 2.4 presents the values for this indicator. The median value for DUs outside MENA is about 50 percent greater than in MENA, suggesting that MENA's utilities are overstaffed from a purely technical perspective. Among both the worst performers and the best performers, non-MENA utilities outperform those in MENA by a similar factor, suggesting that overstaffing is a regionwide phenomenon (there are very few observations for the

Table 2.4 Residential Connections per Full-Time Equivalent Employee for MENA and Non-MENA Utilities, 2013 (or most recent year with data, 2009–12)

Utility type	Region	Number of utilities	Quartile 1: worst performers	Median	Quartile 3: best performers
All	MENA	24	124	184	311
	Non-MENA	68	225	336	654
Distribution	MENA	19	151	252	317
	Non-MENA	57	272	367	691
Vertically integrated	MENA	5	44	90	173
	Non-MENA	11	102	157	284

Source: World Bank calculations.
Note: MENA = Middle East and North Africa.

MENA VIUs and, accordingly, results for this subgroup should be treated with caution).

Figure 2.3a shows the values for residential connections per employee among DUs in MENA. REDAL in Morocco has the highest value by far, and this is well up in the range of best performers outside MENA. At the other extreme, the worst performer—Tubas, in the West Bank—has an exceptionally low value. This may well be due to a social policy to increase employment even when this may drive up costs.

Figure 2.3b plots the values for residential connections per employee for VIUs in MENA. The values for Saudi Electricity Company (SEC) in Saudi Arabia and Dhofar Power Company (DPC) in Oman are above the MENA and non-MENA medians and suggest that these utilities are appropriately staffed. The value for RAECO in Oman is very low, which is consistent with the fact that this is the only rural VIU in our sample.

Distribution losses (percent). Higher losses indicate lower efficiency levels. Because there is no a priori reason that the losses should be different between DUs and VIUs, both types of utility can be directly compared. Table 2.5 shows the quartile data for distribution losses (percent) inside and outside MENA. The median MENA value is slightly lower than the non-MENA value. The best-performing group in MENA is also slightly better than the equivalent non-MENA group.

Figure 2.3 Residential Connections per Full-Time Equivalent Employee for Distribution and Vertically Integrated Utilities, MENA, 2013 (or most recent year with data, 2009–12)

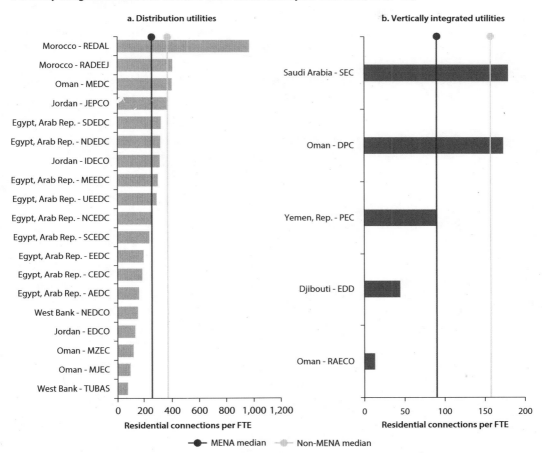

Source: MENA Electricity Database and World Bank calculations.
Note: FTE = full-time employee; MENA = Middle East and North Africa.

Table 2.5 Distribution Losses in MENA and Non-MENA Utilities, 2013 (or most recent year with data, 2009–12)

Region	Number of utilities	Quartile 1: best performers (%)	Median (%)	Quartile 3: worst performers (%)
Non-MENA	114	9	12	18
MENA	37	8	11	14

Source: World Bank calculations.
Note: MENA = Middle East and North Africa.

Figure 2.4 illustrates the performance of the DUs and VIUs in the MENA region. The performance of the VIUs is generally poor. The worst cases—the Ministry of Electricity (MOE) in Iraq, the Public Electricity Corporation (PEC) in the Republic of Yemen, EdL in Lebanon, and JDECO in the West Bank—have losses far above the median values for both MENA and non-MENA, and even the

Figure 2.4 Distribution Losses of Distribution Utilities and Vertically Integrated Utilities in MENA (%), 2013 (or most recent year with data, 2009–12)

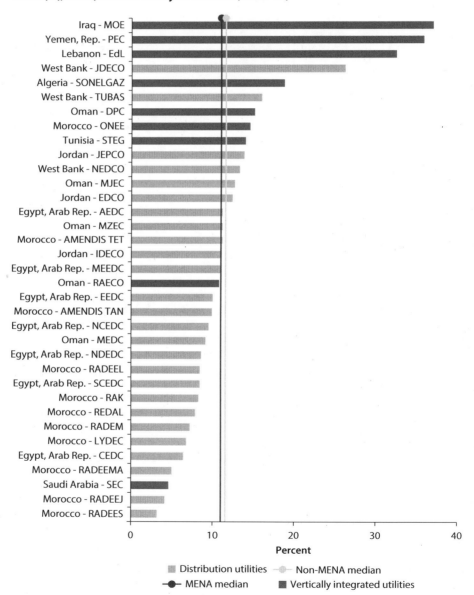

Source: MENA Electricity Database and World Bank calculations.
Note: MENA = Middle East and North Africa.

value of the third quartile for non-MENA countries. This is a variable that can be tackled directly and quickly and is an obvious target for any efforts to improve utility performance. On the lower end, most Moroccan DUs and SEC in Saudi Arabia appear to be good performers.

Commercial Indicators

Energy volume sold per connection. This indicator relates to the scale of operations rather than efficiency and is particularly related to the customer base composition (for example, residential, commercial, industrial). As the economy grows over time, industrialization and household incomes rise and the demand for electricity increases accordingly. If there are economies of scale at any stage of production, then average costs of supply fall, and this can be interpreted as a form of efficiency gain. Table 2.6 shows the values of energy sold per connection. Considering all DUs and VIUs, MENA median sales are larger than non-MENA, as are the sales of the Q1 group. However, for the Q3 group (the best performers), values are similar in both MENA and non-MENA.

Table 2.6 Volume of Energy Sold per Connection for MENA and Non-MENA Utilities, 2013 (or most recent year with data, 2009–12)

Region	Number of utilities	Quartile 1: worst performers (kWh)	Median (kWh)	Quartile 3: best performers (kWh)
Non-MENA	133	2,103	3,405	5,730
MENA	35	3,551	4,223	5,724

Source: World Bank calculations.
Note: kWh = kilowatt-hours; MENA = Middle East and North Africa.

Figure 2.5 illustrates the performance of the MENA DUs and VIUs with respect to energy sales per connection. Three VIUs—SEC in Saudi Arabia and DPC and RAECO in Oman—have values far greater than other VIUs or DUs; these are also higher than the median values both inside MENA and outside. This reflects the relatively high income levels of the consumers served by these utilities.

Total billing per connection ($). This indicator measures scale effects: a higher ratio of billing to connections suggests that the utility's operations are on a relatively sustainable path. Again, an important exogenous factor is the composition of the customer base. Everything else being equal, higher tariffs should be associated with higher billing per connection. There is no a priori reason to expect values to be different between VIUs and DUs. Table 2.7 lists the values for billing per connection. MENA utilities in the median quintile perform slightly better than non-MENA utilities and substantially worse in Q1 and Q3.

The spread of values across both DUs and VIUs in the MENA region is shown in figure 2.6. RAECO in Oman performs very well, with a value above

Figure 2.5 Energy Sales Volume per Connection for Distribution and Vertically Integrated Utilities in MENA (kWh), 2013 (or most recent year with data, 2009–12)

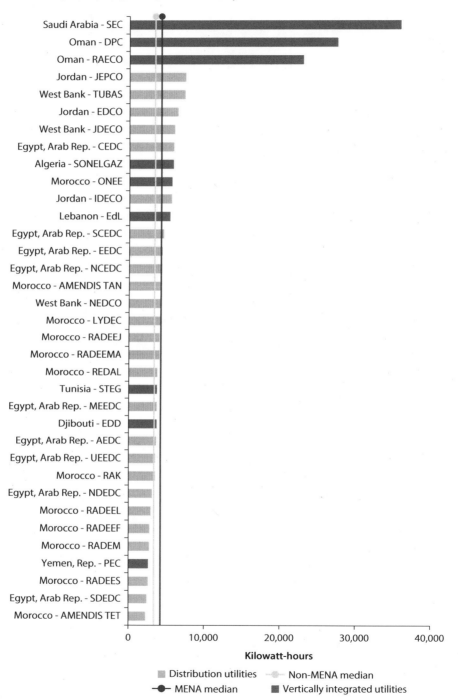

Source: MENA Electricity Database and World Bank calculations.
Note: kWh = kilowatt-hours; MENA = Middle East and North Africa.

Table 2.7 Total Billing per Connection for MENA and Non-MENA Utilities, 2013 (or most recent year with data, 2009–12)

Region	Number of utilities	Quartile 1: worst performers ($)	Median ($)	Quartile 3: best performers ($)
Non-MENA	72	199	292	531
MENA	27	135	299	439

Source: World Bank calculations.
Note: MENA = Middle East and North Africa.

Figure 2.6 Total Billing per Connection for Distribution and Vertically Integrated Utilities in MENA ($), 2013 (or most recent year with data, 2009–12)

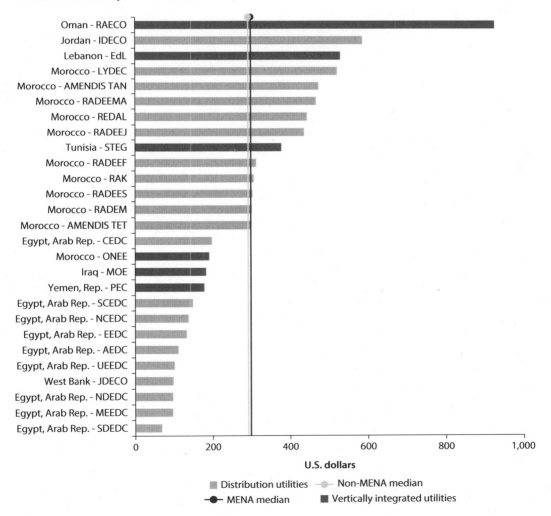

Source: MENA Electricity Database and World Bank calculations.
Note: MENA = Middle East and North Africa.

Shedding Light on Electricity Utilities in the Middle East and North Africa
http://dx.doi.org/10.1596/978-1-4648-1182-1

the non-MENA Q3. Most of Egypt's DUs have very low values, for reasons that should be explored.

Collection rate (percent). Failure to collect the total amount due is an important source of inefficiency, because it leads to deficits and the underfunding of future investment needs. Table 2.8 lists the collection rates of utilities inside and outside MENA. It is expected that DUs and VIUs should be able to achieve similar collection rates. The performance of MENA and non-MENA utilities is very similar at the median, Q1, and Q3 values. It might be noted—and it is surprising, given the importance of this indicator for

Table 2.8 Collection Rates for MENA and Non-MENA Utilities, 2013 (or most recent year with data, 2009–12)

Region	Number of utilities	Quartile 1: worst performers (%)	Median (%)	Quartile 3: best performers (%)
Non-MENA	15	88	94	97
MENA	15	85	92	94

Source: World Bank calculations.
Note: MENA = Middle East and North Africa.

overall efficiency—that only 15 MENA utilities shared the value for this key indicator.

Insights into the poor performance of the weakest MENA utilities are offered by figure 2.7. Most VIUs and DUs have collection rates near the MENA median, but RAECO (Oman), Tubas (West Bank), and especially EDD (Djibouti) have very low collection rates. This points to clear weaknesses in these utilities.

Financial Indicators

OPEX recovery from sales (percent). This indicator measures the extent to which a utility is recovering operating expenditures from its sales of energy. Higher values indicate better performance, and it is interesting to note if the coverage is greater than 100 percent (full recovery). Table 2.9 presents the results for MENA and non-MENA utilities. DUs and VIUs cannot be compared directly because OPEX cannot be disaggregated by individual function using the MED. The MENA median value for DUs (93 percent) is below that for non-MENA (98 percent), but MENA VIUs have a higher median value than non-MENA VIUs. However, Q3 VIUs in MENA perform worse than those outside. This result may reflect the geographical composition of the non-MENA utilities included in the analysis: most of the VIUs are in Sub-Saharan Africa, whereas the DUs are largely in Latin America. On the other hand, DUs in Q1 appear to be doing slightly better in MENA than beyond.

Figure 2.7 Collection Rates for Distribution and Vertically Integrated Utilities in MENA (%), 2013 (or most recent year with data, 2009–12)

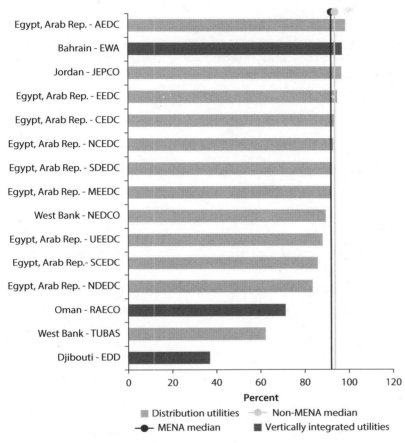

Source: MENA Electricity Database and World Bank calculations.
Note: MENA = Middle East and North Africa.

Table 2.9 OPEX Recovery as a Share of Sales (%) for MENA and Non-MENA Utilities, 2013 (or most recent year with data, 2009–12)

Utility type	Region	Number of utilities	Quartile 1: worst performers (%)	Median (%)	Quartile 3: best performers (%)
All	MENA	32	85	92	99
	Non-MENA	21	77	87	106
Distribution	MENA	23	87	93	99
	Non-MENA	5	81	98	103
Vertically integrated	MENA	9	55	92	99
	Non-MENA	16	77	87	109

Source: World Bank calculations.
Note: MENA = Middle East and North Africa; OPEX = operating expenses.

Shedding Light on Electricity Utilities in the Middle East and North Africa
http://dx.doi.org/10.1596/978-1-4648-1182-1

Figure 2.8a focuses on DUs in MENA and presents the values for OPEX recovered from energy sales. Utilities in Morocco and Jordan appear to perform relatively well on this indicator, whereas those in Egypt and Oman are at the other end of the spectrum. But these results should be considered with caution. The surveys that feed the MED define energy sales as actual sales, without government transfers. However, some utilities seem to have included the government transfers in their responses. This makes comparison difficult and could explain differences across utilities. For example, looking at figure 2.8a, all Oman's utilities—MEDC, Majan Electricity Company (MJEC), and MZEC—are well below the MENA median and, importantly, the breakeven point of 100 percent. But if government subsidies are included, these same three utilities' OPEX recovery values rise to 117 percent, 118 percent, and 126 percent, respectively.

Figure 2.8b presents OPEX recovery values for VIUs in MENA. The value of EDD in Djibouti is well above that of other utilities, whereas EdL in Lebanon is well below, for reasons that require further research to understand. The low value

Figure 2.8 OPEX Recovery from Sales for Distribution and Vertically Integrated Utilities, MENA (%), 2013 (or most recent year with data, 2009–12)

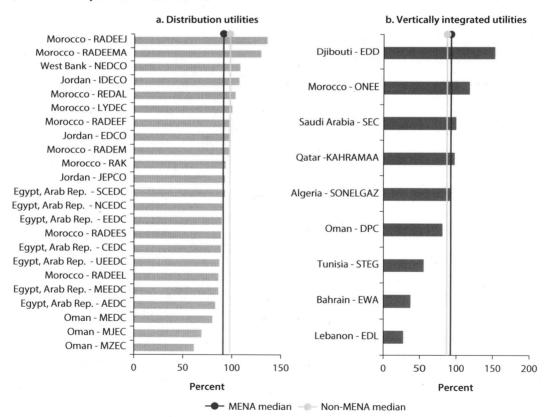

Source: MENA Electricity Database and World Bank calculations.
Note: MENA = Middle East and North Africa; OPEX = operating expenses.

for Bahrain's Electricity and Water Authority (EWA) can be explained by the fact that it did not include government transfers.

Energy sales as a share of total costs (percent). This indicator is a direct measure of a utility's ability to cover all its costs. Values less than 100 percent indicate that total costs are not being recovered for various reasons, all of which can be described as inefficiency. DUs and VIUs cannot be directly compared because the distribution costs of VIUs cannot be separated out from other cost categories. Table 2.10 presents results for MENA and non-MENA utilities. The sample size of non-MENA utilities is notably small: 8 in all categories. The MENA sample of 19 is also small when compared with those used for other indicators. At 67 percent, the median value for non-MENA DUs is well below the cost-recovery level; MENA's is closer, at 88 percentage points. It should be noted, however, that there were only two non-MENA observations, so the values for non-MENA utilities could be underestimates. The VIUs inside MENA are less efficient than those outside MENA, as can be observed by their median values.

Table 2.10 Energy Sales as a Share of Total Costs (%) for MENA and Non-MENA Utilities, 2013 (or most recent year with data, 2009–12)

Utility type	Region	Number of utilities	Quartile 1: share of worst performers (%)	Median (%)	Quartile 3: share of best performers (%)
All	MENA	19	74	87	91
	Non-MENA	8	45	67	79
Distribution	MENA	12	82	88	93
	Non-MENA	2	54	67	79
Vertically integrated	MENA	7	43	56	80
	Non-MENA	6	50	67	73

Source: World Bank calculations.
Note: MENA = Middle East and North Africa.

The performance of individual DUs in MENA is shown in figure 2.9a. Only the Northern Electricity Distribution Company (NEDCO) in the West Bank and the El Jadida municipal utility (RADEEJ) in Morocco have a sales-to-total-cost ratio greater than 1. Moroccan DUs tend to perform better, whereas the majority of Egyptian utilities are the furthest from total cost recovery. Figure 2.9b shows the performance of VIUs in MENA. EDD in Djibouti is the only utility covering total costs, whereas SEC in Saudi Arabia and EdL in Lebanon cover only a small fraction of total costs.

Accounts receivable to sales (days). This indicator measures the time it would take, at current sales levels, to collect all bills. It is used to estimate the number of times a utility is able to convert its credit sales to cash during a financial year.

Figure 2.9 Sales as a Share of Total Costs for Distribution and Vertically Integrated Utilities, MENA (%), 2013 (or most recent year with data, 2009–12)

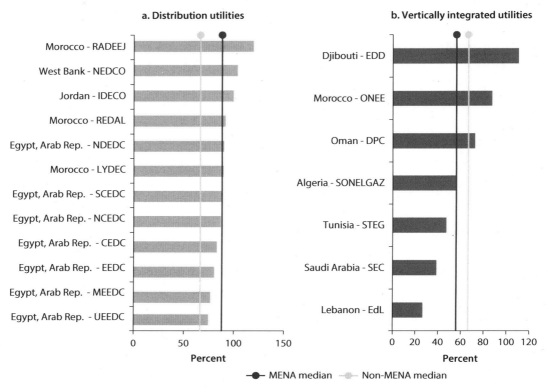

Source: MENA Electricity Database (MED) and World Bank calculations.
Note: MENA = Middle East and North Africa.

Table 2.11 Ratio of Accounts Receivable to Sales in MENA and Non-MENA Utilities, 2013 (or most recent year with data, 2009–12)

Region	Number of utilities	Quartile 1: best performers (days)	Median (days)	Quartile 3: worst performers (days)
Non-MENA	7	8	52	140
MENA	26	117	148	202

Source: World Bank calculations.
Note: MENA = Middle East and North Africa.

The higher the indicator is, the higher collection efficiency and the higher the utility's liquidity value.[3] Table 2.11 presents estimated values inside and outside MENA (where the sample size is very small, at 7). The MENA median value is 148 days, whereas that of non-MENA utilities is only 52 days. If these figures are representative, then it appears that MENA performs relatively poorly on this indicator. The difference is even more stark for the best-performing utilities: the Q1 value for non-MENA utilities is 8.4 days.

The ratios of accounts receivable to sales among MENA utilities are shown in figure 2.10. The large majority have values greater than 100 days. All utilities in MENA have values above the non-MENA median.

Figure 2.10 Accounts Receivable to Sales for Distribution and Vertically Integrated Utilities Utilities in MENA (days), 2013 (or most recent year with data, 2009–12)

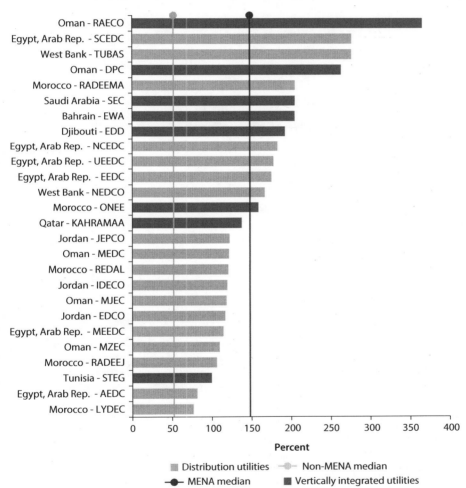

Source: MENA Electricity Database and World Bank calculations.
Note: MENA = Middle East and North Africa.

Debt to equity. A high debt-to-equity ratio indicates an aggressive growth-financing approach to debt. Risks to this approach include the cost of additional interest expenses and any volatility if the debt is short to medium. If the cost of debt financing outweighs the returns generated by the additional capital, the financial load quickly becomes an issue, whether the utility is publicly or privately owned. This is why a common rule of thumb is to cap the ratio at 2.

Table 2.12 presents the values of the ratio for MENA and non-MENA utilities. With no a priori reason to argue that the debt-to-equity ratio should be different across utility types, no distinction is made in this table among VIUs, DUs, GUs, and TUs. The MENA median value is 357 percent whereas the non-MENA median, based on a small sample, is only 91 percent. The very high MENA value suggests an excessive reliance on debt financing.

Table 2.12 Ratio of Debt to Equity for MENA and Non-MENA Utilities, 2013 (or most recent year with data, 2009–12)

Region	Number of utilities	Quartile 1: best performers (%)	Median (%)	Quartile 3: worst performers (%)
Non-MENA	14	62	91	443
MENA	47	207	357	767

Source: World Bank calculations.
Note: MENA = Middle East and North Africa. Data are for all utility types: vertically integrated, distribution, generation, and transmission.

Figure 2.11 shows the spread of MENA values by utility type. The distribution is uneven. Several utilities, in particular GUs in Egypt and Oman, have debt-to-equity ratios over 10:1. Understanding the reasons behind these surprisingly high values would require detailed investigation, though the case studies of Egypt and Oman in part II of this book offer some insights.

In 2013, GUs suffered from an average debt-to-equity ratio of the order of 20:1. Corporate governance should be improved to ensure these utilities' restructuring, with the aim of long-term sustainability. In the meantime, two concrete actions could be taken: first, raise these utilities' equity by converting public debt into equity; and second, reform tariffs (see, for example, the many suggestions made by Egypt's regulator, ERA).

Current assets to current liabilities (percent). The ratio of current assets to current liabilities measures the extent to which short-term assets (cash, cash equivalents, marketable securities, and receivables) are readily available to pay off short-term liabilities (payables, current portion of term debt, accrued expenses, and taxes). Generally, the higher the ratio, the better. The values for all types of MENA and non-MENA utilities are shown in table 2.13. The median

Figure 2.11 Ratio of Debt to Equity across Utility Types in MENA (%), 2013 (or most recent year with data, 2009–12)

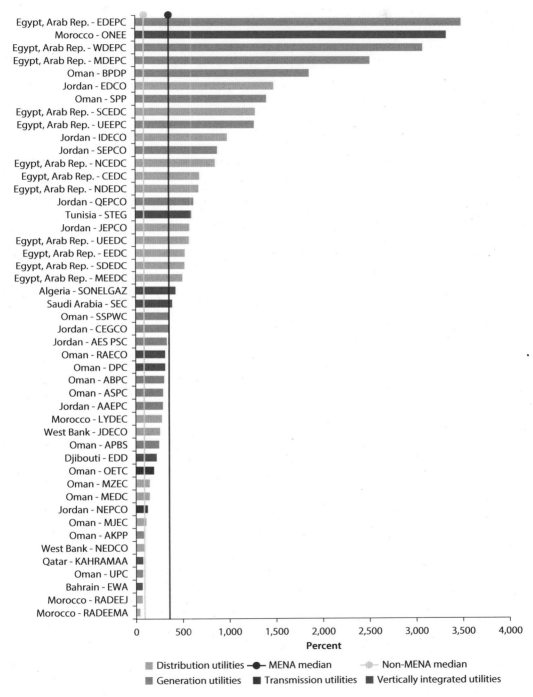

Source: MENA Electricity Database and World Bank calculations.
Note: MENA = Middle East and North Africa.

Shedding Light on Electricity Utilities in the Middle East and North Africa
http://dx.doi.org/10.1596/978-1-4648-1182-1

non-MENA value (88 percent) and the median MENA value (84 percent) are similar, and both raise concerns because they are below 100 percent. The Q1 and Q3 values are also similar both inside and outside MENA.

Table 2.13 Ratio of Current Assets to Current Liabilities for MENA and Non-MENA Utilities, 2013 (or most recent year with data, 2009–12)

Region	Number of utilities	Quartile 1: worst performers (%)	Median (%)	Quartile 3: best performers (%)
Non-MENA	19	62	88	119
MENA	53	63	84	121

Source: World Bank calculations.
Note: MENA = Middle East and North Africa.

Figure 2.12 plots the ratio of current assets to current liabilities across all types of utilities in MENA. Several GUs—the Wadi Al Jizzi Power Company (WAJPCO) in Oman, Qatrana Electric Power Company (QEPCO) in Jordan, Al-Ghubra Power & Desalination Company (GPDCO) in Oman, and Amman East Power Plant (AES PSC) in Jordan—boast a ratio of 3:1, indicating a strong financial position. On the other hand, a majority of utilities (33 out of 53) falls below the 1:1 threshold, and nine have a ratio of less than 1:2, indicating a weak financial position.

Return on assets (percent). ROA is a measure of profitability. The higher the value, the better the performance. However, comparisons across markets need to recognize that riskier markets will require higher returns. Table 2.14 presents the results for MENA and non-MENA utilities. The sample sizes are very small for all the non-MENA categories, but some comparisons can be made. Across all utilities, the median ROA for the non-MENA group is 1 percent, whereas for MENA it is 3 percent. Although the same pattern is observed for Q1, in Q3 MENA utilities appear to have lower ROA than non-MENA ones. All these values are low and suggest generally weak financial performance.

Figure 2.13 illustrates the ratios of individual VIUs in MENA. The value of Lebanon's VIU, EdL, which has an ROA value of −150 percent, is not represented. The Jordan Electric Power Company (JEPCO), and Morroco's Régie

Table 2.14 Return on Assets for MENA and Non-MENA Utilities, 2013 (or most recent year with data, 2009–12)

Region	Number of utilities	Quartile 1: worst performers (%)	Median (%)	Quartile 3: best performers (%)
Non-MENA	12	−2	1	9
MENA	49	0	3	6

Source: World Bank calculations.
Note: MENA = Middle East and North Africa.

Figure 2.12 Ratio of Current Assets to Current Liabilities: Selected Utilities of All Types, MENA (%), 2013 (or most recent year with data, 2009–12)

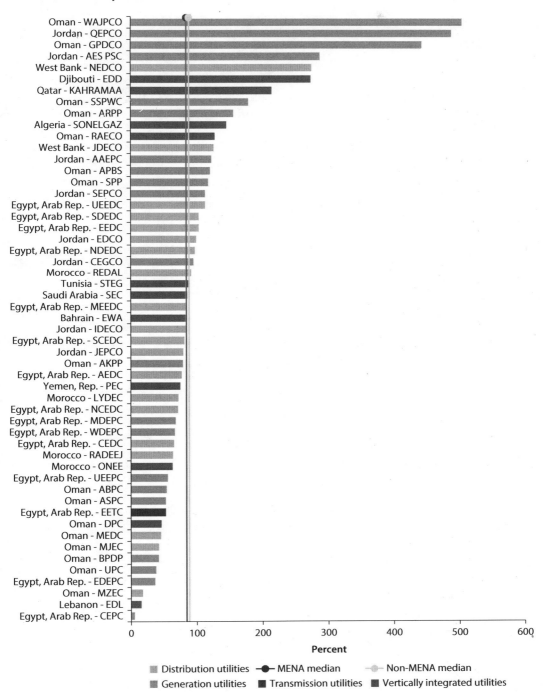

Source: MENA Electricity Database and World Bank calculations.
Note: MENA = Middle East and North Africa.

Shedding Light on Electricity Utilities in the Middle East and North Africa
http://dx.doi.org/10.1596/978-1-4648-1182-1

Autonome de Distribution d'Eau et d'Électricité de Meknès (RADEM) and
Régie Autonome Intercommunale de Distribution d'Eau et d'Électricité de Safi
(RADEES) have the highest ROAs, with values in excess of 10 percent. At the
low end, JDECO (West Bank) stands out with a very large negative return.
This needs to be investigated further to understand any special factors.
AMENDIS TET (Morocco), STEG (Tunisia), Sociète Nationale de l'Electricité
et du Gaz (SONELGAZ) (Algeria), and Office National de l'Electricité et de
l'Eau Potable (ONEE) (Morocco) also have negative ROAs. It appears that VIUs
tend to be at the lower end of the distribution, with almost all of them below the
MENA median.

Return on equity (ROE) (percent). This indicator measures the return on
shareholders' investments. Table 2.15 compares the values for MENA and non-
MENA utilities. Again there are few observations for the non-MENA economies.
The median MENA value of 6 percent is well above the non-MENA value of
0 percent, and there are similar differences in performance between the Q3 and
Q1 utilities.

Figure 2.14 plots the values for the individual MENA utilities. The value of
Morocco's VIU, ONEE, which has a ROE value of −127 percent, is not represented. The top seven performers achieved an ROE of 10 percent or better,

**Table 2.15 Return on Equity for MENA and Non-MENA Utilities, 2013 (or most recent
year with data, 2009–12)**

Region	Number of utilities	Quartile 1: worst performers (%)	Median (%)	Quartile 3: best performers (%)
Non-MENA	13	−4	0	6
MENA	46	0	6	16

Source: World Bank calculations.
Note: MENA = Middle East and North Africa.

whereas four utilities had negative values. The markedly negative values for
STEG (Tunisia) and JDECO (West Bank) require further research. GUs appear
to be at the upper end of the ROE DUs, and VIUs, once again, are at the
lower end.

Figure 2.13 Return on Assets: Selected Utilities of All Types, MENA (%), 2013 (or most recent year with data, 2009–12)

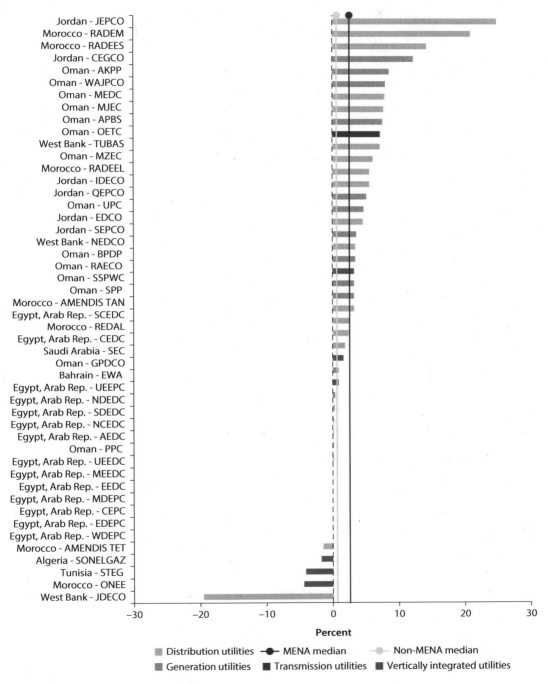

Source: MENA Electricity Database and World Bank calculations.

Note: MENA = Middle East and North Africa. The value of EdL (Lebanon) which has a value of −150% is not represented on this graph for a matter of scale and representation.

Shedding Light on Electricity Utilities in the Middle East and North Africa
http://dx.doi.org/10.1596/978-1-4648-1182-1

Figure 2.14 Return on Equity for Selected Utilities of All Types in MENA (%), 2013 (or most recent year with data, 2009–12)

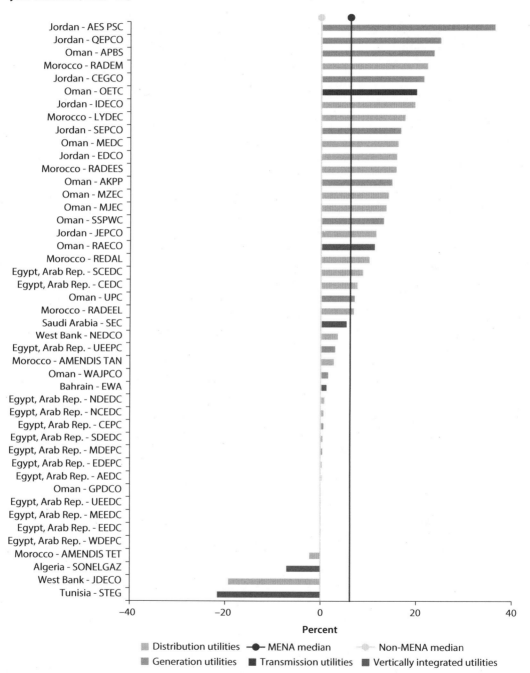

Source: MENA Electricity Database and World Bank calculations.

Note: MENA = Middle East and North Africa. The value of ONEE (Morocco) which has a value of −127% is not represented on this graph for a matter of scale and representation.

Conclusion

Calculating the median values of various performance indicators allows a comparison between MENA and non-MENA electricity utilities. Despite the challenges of data availability and comparability, it has been possible to indicate where the utilities of the MENA region fall behind—or exceed—the performance of comparable utilities elsewhere.

When the values of indicators for individual utilities in MENA are plotted against the MENA and non-MENA medians, some valuable insights can be gained. This approach allows outlying values to be identified so that further research can be focused on these cases. It also highlights indicators for which there is relatively little variation across utilities in the MENA region and those for which there is a large gap between best and worst performers (even after excluding clear outliers). Policy makers concerned with the performance of individual utilities in their economies can use these tools to set realistic targets for improvement and can monitor progress toward specific objectives.

One interesting feature of the comparisons within MENA is that there were no immediately obvious "best" or "worst" performers among utilities across all indicators. An approach to measuring performance across several indicators is presented in chapter 4.

Notes

1. The non-MENA data were drawn from 38 vertically integrated utilities, 4 generation utilities, 135 distribution utilities, and 4 transmission utilities that are found in appendix B. The list of non-MENA utilities can also be found in appendix B.

2. All indicator values expressed in monetary terms are converted to U.S. dollars at survey year exchange rates. For intercountry comparisons, valuations at purchasing power parity (PPP) could well affect the relative magnitudes of the MENA and non-MENA indicators.

3. The average collection period is computed as the number of days in that period divided by the accounts-receivable-to-sales ratio. A value of 6 means that the average client pays once every two months. This implies an average collection period of 60 days. Shortening the collection period reduces the working capital cycle and often eases access to bank loans for needed investments.

CHAPTER 3

A Dynamic Look at MENA Performance over Five Years

Comparing performance indicators over time is of interest when assessing whether utilities are improving their efficiency or not. Where governments have introduced power sector reforms, policy makers might examine the effects using certain indicators. Changes are expected to be gradual rather than sudden and may take several years to see. Other factors, apart from government policy, may impact indicators. Local demand, the international economic climate (including oil price fluctuations, for example), and political issues can all influence utility performance.

Data Challenges

In the survey that informs the Middle East and North Africa (MENA) Electricity Database, utilities were asked for information on several years, from 2009 to 2014, not just the base year (2013). Given that less data were obtained for 2014 than for 2009–13, we have not included that year in the analysis summarized in most of this book. This chapter is an exception, given its focus on the dynamic aspects of performance. The large number of indicators included in the MENA Electricity Database (36 core indicators) and the large number of utilities surveyed (67) meant that considering data at a utility level would require nearly 3,000 trend calculations to be made. When the data series are so short, and with inevitable questions concerning data accuracy, such an exercise would not be sensible. An alternative is to consider constructing aggregates across utilities, indicator by indicator, and to carry out trend analysis on these aggregates for the few years of data available.

Preliminary examination of the data revealed that the coverage of each indicator is only partial, even for the base year, and utilities supplied data for different years within the six-year span requested. Few provided information for all 2009–14. This meant that it was not possible to set up a standard comparison between indicators or between utilities.

The first step in the analysis was to compare average indicator values for each year based on (a) only those utilities that provided data for at least five of the six years (see table 3.1) and (b) all utilities answering for that year, this number varying from year to year (see table 3.2). This comparison was made for four groups of utilities: all utilities, vertically integrated utilities (VIUs), distribution utilities (DUs), and generation utilities (GUs). The median values were then compared within each group to avoid the impacts of extreme values caused by issues in data collection, which might exist for only one year.

To carry out this analysis we started concentrating on a single indicator of substantial importance to performance: the ratio of current assets to current liabilities, which was collected for at least five years by 35 utilities (seven VIUs, 15 GUs, and 13 DUs). The median values for each year, by utility type, based on these respondents (the common sample) are shown in table 3.1. The median values, based on all respondents for each year (that is, more than 35 utilities for each year) are shown in table 3.2.

Tables 3.1 and 3.2 show a significant decline in most values as of 2014, which is caused by the sudden change in the number of utilities reporting for that year. The smaller sample in 2014 consisted of utilities that tended to have the lowest values in the sample, thus pulling the average down. Thus, 2014 should be excluded from any analysis in which the number of reporting utilities is notably smaller than for the rest of the years. Using all available data points on each year for time series analysis would produce movements that are due largely to the inclusion or exclusion of some utilities (that is, those who provided information for only four years or fewer).[1] In addition, one utility's performance on a given

Table 3.1 Median Values of Ratio of Current Assets to Current Liabilities for Utilities (%), 2009–14

Minimum of five observations

Type of utility	2009	2010	2011	2012	2013	2014
All	95	100	92	82	84	47
Vertically integrated	125	105	119	75	89	43
Generation	107	123	127	118	113	47
Distribution	93	97	89	91	80	37

Source: World Bank calculations.
Note: The median samples are based on a common sample.

Table 3.2 Median Values of Ratio of Current Assets to Current Liabilities for Utilities (%), 2009–14

Type of utility	2009	2010	2011	2012	2013	2014
All	95	105	96	87	85	52
Vertically integrated	111	105	96	75	109	76
Generation	93	100	100	107	87	105
Distribution	89	98	86	83	81	39

Source: World Bank calculations.
Note: Observations not necessarily available for all utilities for a given year.

indicator could fluctuate widely over time, possibly due to problems of data collection and recording.

Indicator Trends with All Utilities Aggregated

Our preliminary findings led us to begin with sets of common data and analyze the behavior of the median between 2009 and 2014, measured over all utilities in the common dataset for one indicator. With this dataset, tests were carried out to look for the existence of a significant trend by regressing the log of the indicator value in each year on a time trend variable (increasing by one unit each year). If the coefficient of the trend term was not significantly different from zero, then we concluded there was no trend in the indicator, so it effectively remained constant.

Table 3.3 estimates the trend value and the probability of significance of the 25 indicators for which there were adequate data, meaning that the proportion of missing observations for 2014 relative to the number of utilities was small.

Table 3.3　Estimated Trend of Indicators for Utilities, 2009–14

Indicator	Number of utilities with common data	Estimated trend value	Probability
Availability factor	6	0.004	0.34
Capacity factor	17	0.010	0.57
Load factor	15	−0.002	0.73
Percentage of meters replaced	11	−0.110	0.12
Network maintenance	12	−0.070	0.49
OPEX/employee	2	0.100	0.15
OPEX/connection	26	0.150	0.04*
OPEX/km	29	−0.010	0.74
Residential connections/employee	13	−0.150	0.12
Sales/employee	24	−0.020	0.69
Revenue/employee	25	−0.010	0.69
Fuel/OPEX	15	0.010	0.84
Energy purchase/OPEX	35	0.010	0.29
Labor costs/OPEX	26	−0.060	0.01*
Sales/OPEX	38	−0.020	0.01*
Accounts receivable/sales	34	0.020	0.52
Debt/equity	38	0.050	0.11
Current assets/current liabilities	35	−0.040	0.08
Return on assets	39	−0.040	0.70
Return on equity	42	0.090	0.29
Total billing/connection	19	0.110	0.09
Collection rate	15	0.010	0.55
Prepaid meters installed (%)	6	0.250	0.10
Distribution losses	27	0.040	0.22
SAIFI	10	0.020	0.51

Source: World Bank calculations.
Note: km = kilometer; OPEX = operating expenses; SAIFI = System Average Interruption Frequency Index.
Significance level: * = 5%.

Shedding Light on Electricity Utilities in the Middle East and North Africa
http://dx.doi.org/10.1596/978-1-4648-1182-1

Using a two-sided test, which would accept evidence for either an increasing or a decreasing trend based on five years of data, the probability should be less than 0.05 to reject the hypothesis of no trend in the data.

Three indicators showed significant time trends during the period studied, when measured against aggregate data (median values): (a) operating expenses (OPEX) per connection, which had a 15 percent annual growth rate; (b) labor costs to OPEX, with a negative annual growth rate of 6 percent; and (c) sales to OPEX, with a negative annual growth rate of 2 percent. These figures suggest that OPEX increased significantly throughout the region between 2009 and 2014, although the number of connections increased only slowly, and sales and labor costs increased at a moderate rate. Because the main components of OPEX are fuel and labor costs, the very large increase in oil prices at the beginning of the period[2] is likely to have influenced these trends. The increase in the value of OPEX might also be attributable to an increase in wages or to a renewed emphasis on maintenance and repair. A detailed breakdown of OPEX would be needed to understand the reasons for these trends.

The growth rates estimated for the highly aggregated data are nearly all zero, which suggests that there are few regionwide trends that can be identified with such a short run of data.

Indicator Trends Disaggregated by Utility Type

Some patterns may be observable when a more disaggregated approach is used. This is illustrated by the ratio of current assets to current liabilities. Table 3.4 contains the estimated growth rates for the indicator based on data from utilities with a common sample of all observations from 2009 to 2013. Median values are constructed for VIUs, DUs, and GUs. The table shows that the groups produced different values of the estimated trend, but only the generators produce a significant positive annual trend of 8 percent. Therefore, we note that trends can differ among types of utilities for reasons connected with their nature.

The same analysis was undertaken for the remaining indicators from table 3.3.[3] Four additional disaggregated trends were found to be significant for a particular type of utility (table 3.5).

Table 3.4 Estimated Trends for Median Ratio of Current Assets to Current Liabilities, by Utility Type, 2009–13

Utility type	Number of observations	Trend value	Probability
All	35	−0.04	0.08
Vertically integrated	7	−0.10	0.14
Distribution	13	−0.03	0.11
Generation	15	0.08	0.01*

Source: World Bank calculations.
Note: The estimated trends were based on a common sample.
Significance level: * = 5%.

Table 3.5 Estimated Trends, by Utility Type, 2009–13

Utility type	Number of observations	Trend value	Probability
a. Median capacity factor			
All	17	0.014	0.570
Vertically integrated	3	−0.390	0.040*
Generation	14	0.012	0.570
b. Median OPEX per employee			
All	24	0.110	0.150
Vertically integrated	7	0.087	0.520
Distribution	7	−0.180	0.120
Generation	8	0.470	0.002*
c. Median sales per employee			
All	24	−0.017	0.690
Vertically integrated	6	0.220	0.090
Distribution	7	−0.260	0.070
Generation	9	0.380	0.020*
d. Return on assets			
All	39	−0.047	0.700
Vertically integrated	6	−0.009[a]	0.010*
Distribution	18	−0.260	0.140
Generation	11	−0.018	0.790

Source: World Bank calculations.
Note: The estimated trends were based on a common sample. OPEX = operating expenses.
Significance level: * = 5%.
a. The number is a linear trend because of negative values for vertically integrated utilities.

The capacity factor of VIUs appears to be declining over time, but caution should be taken when interpreting this result given the very small sample size. For GUs, OPEX per employee grew 47 percent a year—probably driven by the very large increase in oil prices at the beginning of the period—and total sales (in monetary value) per employee grew by 38 percent a year. This could reflect the pass-through of oil price variations from generators to the VIUs or the transmission utilities (TUs) buying their electricity, translated into increased sales due to the increase in oil prices during the period of interest. Finally, the return on assets (ROA) of VIUs appears to have decreased slightly over time (by 0.9 percent).

To further disaggregate data for trend analysis, we use the ratio of current assets to current liabilities for the large group of nine GUs in Oman that all provided full data for this indicator (2009–13). This group should be free of the large intercountry differences caused by differences in policies and economic conditions, allowing common trends to be identified. Table 3.6 presents the results of a trend analysis carried out for each utility and for the median of the group.

The results in table 3.6 indicate that for only one of the GUs (the Al-Ghubra Power & Desalination Plant, GPDCO) was there a significant trend in the ratio of current assets to current liabilities: an estimated decline of 27 percent per year.

Shedding Light on Electricity Utilities in the Middle East and North Africa
http://dx.doi.org/10.1596/978-1-4648-1182-1

Table 3.6 Estimated Trends for Ratio of Current Assets to Current Liabilities for Generation Utilities in Oman, 2009–13

Utility	Trend value	Probability
ACWA Power Plant	0.130	0.21
Al-Ghubra Power & Desalination Plant	−0.270	0.01*
Al-Kamil Power Plant	−0.070	0.20
Al-Rusail Power Plant	0.060	0.08
Barka Power and Desalination Plant	−0.160	0.25
Sembcorp Salalah Power and Water Co.	0.150	0.74
Sohar Power Plant	−0.001	0.99
United Power Company	−0.210	0.09
Wadi Al-Jizzi Power Plant	−0.100	0.41
All generation utilities	0.080	0.01*

Source: World Bank calculations.
Note: The estimated trends were based on a common sample.
Significance level: * = 5%.

Such a steep decline suggests that there must have been special circumstances behind the data, requiring further investigation.

The other nine utilities have insignificant trends, but there is a significant positive trend for the aggregate of all GUs in Oman. This must be affected by the rapid change in this indicator's values for some of the generation utilities in Oman.

Conclusion

The scale of the survey, covering 67 utilities and 36 indicators, prohibited a trend analysis at the utility level. Furthermore, the small number of years for which data were collected (a sample of typically four years within the period 2009–13) implies that trends had to be very well marked to be statistically significant and that errors in data collection could negate apparent results.

As an alternate to trend analysis utility by utility, aggregation across utilities was explored, in which the average for a group of utilities for each year was then subjected to trend analysis. It was shown that the averages should be based on the same utilities in each year (the common sample) because averages based on all data available for each year were very sensitive to gaps in data for some utilities in some years. Trend growth rates were fitted to each indicator for which there were adequate data, based on median values for those utilities in the common sample (that is, that provided information for all years). Only three indicators exhibited significant trends: OPEX per connection exhibited a large positive trend, whereas energy sales over OPEX and labor costs over OPEX exhibited negative trends. These were likely due to the dramatic increase in oil prices during the period.

The level of aggregation used for these tests allowed a small-scale investigation, but too much aggregation can conceal common trends between subgroups. A single indicator (current assets to current liabilities) was used to explore the effects of disaggregation. Disaggregating to the level of utility type (vertically integrated, distribution, generation) within the MENA region revealed a significant positive trend for generators as a group, although the other groups did not have significant trends. This result adds support to the idea that trend analysis, where there are sufficient data, should be carried out for the different types of utilities. We then undertook the same exercise for all the other indicators, and found statistically significant results for the capacity factor, OPEX per employee, sales per employee, and ROA.

Further disaggregation to the level of individual GUs in Oman was carried out for a set of nine such utilities. Conditions within Oman were similar for all utilities, and policies applied equally. One utility exhibited a significant negative trend, so large as to require further investigation. The other eight utilities showed no significant trends, and it was clear that the data series were generally too short to pick up trends in performance. Aggregation, aimed at smoothing out random shocks, also did not reveal much in the way of regionwide trends.

This analysis of the performance of MENA utilities suggests that to identify underlying trends in performance (if any), substantially longer time series of data would be required. Also, analysis should be carried out at a utility level or for aggregates of utilities over all the years analyzed. Taking yearly averages over a varying number of utilities is likely to produce large swings in the aggregate due to its composition.

Notes

1. A similar analysis was carried out using the means rather than medians of the current assets to current liabilities ratios. The same conclusions were reached—and indeed reinforced by noting that changing the size of the sample between years allowed certain extreme observations to dominate the data and produce large fluctuations between years.

2. In 2009, the average price of Brent crude oil was $62 a barrel (bbl); in 2010, it was $80 per bbl; and in 2010, it was $111 per bbl. It fell to $108 per bbl in 2013 and then modestly to $99 per bbl in 2014, which is still 60 percent higher than at the beginning of this period.

3. Only results generating a significant trend for some category of utilities are presented here.

Shedding Light on Electricity Utilities in the Middle East and North Africa
http://dx.doi.org/10.1596/978-1-4648-1182-1

CHAPTER 4

A Multi-Indicator Approach to Analyzing Utility Performance

Chapter 2 ranked 67 utilities across the Middle East and North Africa (MENA) on several indicators and identified utilities whose performance on a particular indicator was very strong or very weak. Understanding the reasons behind these performance levels can inform policies to be applied elsewhere.

Two features of the analysis stand out. First, because of data gaps it was not always possible to compare utilities on more than one indicator. Second, even when a utility provided data on multiple indicators, its performance was not consistently high (or low) across them. The use of a single indicator to assess relative performance offers a somewhat limited view, dominated by the specific characteristics of the utilities surveyed. Some form of average performance measure is required to reveal overall strengths and weaknesses.

To this end, we compared a set of distribution utilities (DUs) against one another, using the same indicators for every utility in the set. This avoided the challenges that necessarily arise when comparing different types of utilities and different indicators.

Methodology

A performance assessment of multiple indicators reflects a wider range of a utility's characteristics and reduces the chance of results being decided by peculiar circumstances. This methodology is useful given that the quality of our data is not sufficient to use more sophisticated approaches (for example, stochastic frontier analysis or data envelopment analysis). The challenge of this multi-indicator approach is combining indicators measured across very different contexts. An average rank score addresses this problem. For example, suppose there are 10 utilities, all of which reported data on their return on equity (ROE) and sales-to-employee ratio. The utility with the highest ROE would receive a score of 10 for that indicator, the next-best utility a score of 9, and so on. The sales-to-employee ratio would be treated similarly. A combined performance measure would then be the average rank value for each utility. This approach can be generalized to include as many indicators as necessary.

The average rank indicator allows indicators based on different measurement units to be combined, but in doing so becomes a purely relative measure of performance. It does not distinguish between large and small actual differences across successive observations, and all indicators are of equal importance. If new data become available for other utilities, they can easily be added into the ranking so new comparisons are possible. An important feature of rank-based indicators is that they are robust against all but very large measurement errors in the original data.

It is interesting to note the extent to which a utility's rank order is similar or not for different indicators. For example, if its ranking is similar for both the ROE and the sales-to-employee ratio, this indicates that the utility tends to be strong or weak across the board. Meanwhile, very different ranks suggest there is little evidence for calling a particular utility strong or weak across all dimensions. The degree of agreement across indicators can be measured using Kendall's coefficient of concordance (W).[1] The maximum value of the coefficient of concordance is unity, and the minimum is zero. At the maximum value, any single indicator would give the same performance ranking as an average of all indicators. The nearer to unity the W coefficient, the less the need to take an average of several indicators to produce an overall ranking of performance. For cases where there are 5 or more utilities or more than 15 indicators, a test of the null hypothesis (that is, that there is no agreement between the rankings of the different indicators) can be carried out.[2]

A high W value indicates that some utilities are more interested in efficiency than others and that they tend to look for improvements in several aspects of performance. If they pursued all avenues toward efficiency with equal effort, but the degree of effort varied among the utilities, then the concordance would be unity, with each utility achieving the same rank for every one of a set of indicators (and different from those of all other utilities). However, if they pursue all avenues toward efficiency with different degrees of effort, the correlations between the ranks will drop and W will move toward zero. In an ideal situation, all utilities would pursue all avenues toward efficiency at the same time and with the same intensity so that W would quickly move toward unity. In practice, it may be easier to focus on just one or two areas of performance, and different utilities may prioritize different target indicators accordingly.

Data Considerations

The issues of data availability and inconsistent ranking can be illustrated using material presented in chapter 2. One DU, the Muscat Electricity Distribution Company (MEDC) in Oman, is used for this purpose, but similar results would have been obtained for other utilities. Table 4.1 gives the rank score for MEDC (over all DUs for which there were data) for all indicators used in the global comparison exercise. Ranking is from the worst performer (1) to the best (value equaling sample size).

Shedding Light on Electricity Utilities in the Middle East and North Africa
http://dx.doi.org/10.1596/978-1-4648-1182-1

Table 4.1 Ranked Performance of MEDC (Oman) on Various Indicators

Indicator	Performance rank	Sample size (value of maximum rank)
OPEX/connection	1	25
OPEX/kWh	—	—
Connections/employee	17	19
Distribution losses	13	28
Sales/connection	—	—
Billing/connection	—	—
Collection rate	—	—
Sales/OPEX	3	23
Sales/costs	—	—
Accounts receivable/sales	11	18
Debt/equity	15	19
Current assets/current liabilities	3	20
Return on assets	13	24
Return on equity	15	24

Source: World Bank calculations.
Note: kWh = kilowatt-hours; MEDC = Muscat Electricity Distribution Company; OPEX = operating expenses; — = not available.

For several of the indicators used for the global comparison there are no data for MEDC, and it cannot be compared to other DUs on these dimensions of performance. For indicators where there are data, the performance is variable. For example, it is high for return on assets (ROA) and ROE, but low for the ratios of assets to liabilities and sales to operating expenses (OPEX). Judging its performance on any single indicator could result in a misleading picture of overall performance, so a measure across several dimensions may be preferred.

The data gaps of the MENA Electricity Database represent significant obstacles to this approach, or any approach based on a number of individual indicators. The database covers a large range of indicators (36) for the 67 utilities, but for many a full set of data was unavailable. The data gaps are different between types of utilities so that the larger the number of indicators considered for the average rank score, the fewer utilities would have data for all the indicators. A balance has to be struck between comparing performance across a large number of utilities (and fewer indicators) and comparing performance across a wide range of indicators to provide a more balanced assessment (using fewer utilities). In addition to data availability, indicators were selected in such a way as to include at least one indicator per performance category (technical, commercial, and financial).

To choose the number of indicators and the set of utilities to include, the data were separated into 12 vertically integrated utilities (VIUs), 23 generation utilities (GUs), 29 DUs, and 3 transmission utilities (TUs). Because 3 is such a small number, we decided to exclude this last group from the average rank score exercise. For each of the other three groups, the utilities with data available on each indicator were identified, as well as those with data for all the most populous indicators in a sequence. This provided a picture of the trade-off between the number of indicators and the number of utilities available for the average rank score exercise.

Distribution Utilities: Average Rank Score

Table 4.2 shows the results of the multi-indicator approach, as applied to DUs. The OPEX per kilometer (km) indicator is the most widely available: 37 observations over all types of utilities and 27 observations for DUs. The total energy volume sold per connection is available for 26 distribution utilities, but only 25 DUs have data on both this and OPEX per km. Add the ratio of energy sales to OPEX, and this limits the set of DUs to 22. As more indicators are added, the number of utilities with data on all drops steadily.

We decided to include the five indicators shown in table 4.2.[3] This group of indicators covers technical, commercial, and financial performance measures, which should help capture different aspects of performance.[4] Table 4.3 lists the individual and average rank scores of each of the 17 DUs for which there were data on all five indicators.

Table 4.2 Trade-Off between Number of Distribution Utilities and Number of Indicators Common to All MENA Utilities

Indicator	All	Distribution utilities only	Common to sequential set of distribution utilities
OPEX/km	37	27	27
Total energy volume sold/connection	35	26	25
Energy sales/OPEX	32	23	22
Return on equity	46	24	18
Revenue/employee	34	26	17

Source: World Bank calculations.
Note: km = kilometer; MENA = Middle East and North Africa; OPEX = operating expenses.

Table 4.3 Ranks and Average Rank Score for Distribution Utilities, MENA

Utility	OPEX/km	Energy volume/ connection	Energy sales/ OPEX	Return on equity	Revenue/ employee	Average rank
Jordan - EDCO	5	16	14	15	12	12.4
Morocco - LYDEC	1	10	15	16	16	11.6
Jordan - JEPCO	3	17	11	13	14	11.6
Morocco - REDAL	2	9	16	12	17	11.2
West Bank - NEDCO	8	11	17	8	10	10.8
Morocco - RADEM	6	3	13	17	13	10.4
Egypt, Arab Rep. - SCEDC	10	14	7	11	9	10.2
Egypt, Arab Rep. - CEDC	12	15	5	10	7	9.8
Egypt, Arab Rep. - NDEDC	15	5	12	7	6	9.0

table continues next page

Table 4.3 Ranks and Average Rank Score for Distribution Utilities, MENA *(continued)*

Utility	OPEX/km	Energy volume/ connection	Energy sales/ OPEX	Return on equity	Revenue/ employee	Average rank
Egypt, Arab Rep. - NCEDC	11	12	8	6	8	9.0
Morocco - RADEES	4	2	9	14	15	8.8
Morocco - RADEEL	7	4	6	9	11	7.4
Egypt, Arab Rep. - UEEDC	16	6	4	3	5	6.8
Egypt, Arab Rep. - EEDC	14	13	3	1	3	6.8
Egypt, Arab Rep. - MEEDC	17	8	2	2	4	6.6
Egypt, Arab Rep. - SDEDC	13	1	10	5	2	6.2
Egypt, Arab Rep. - AEDC	9	7	1	4	1	4.4

Source: World Bank calculations.
Note: km = kilometer; OPEX = operating expenses.

Based on the average rank score, Jordan's Electricity Distribution Company (EDCO) is the best-performing utility in the group, followed by Lyonnaise des Eaux de Casablanca (LYDEC) of Morocco and the Jordan Electric Power Company (JEPCO). EDCO and LYDEC perform very well on four indicators but poorly on OPEX/km, whereas the Alexandria Electricity Distribution Co. (AEDC) in the Arab Republic of Egypt performs very poorly on three indicators but is in the middle of the group for the other two. These examples illustrate the potential danger of relying on a single indicator to describe a utility's performance. AEDC is the worst performer overall, followed by the South Delta Electricity Distribution Company (SDEDC) and the Middle Egypt Electricity Distribution Company (MEEDC), both also in Egypt. It is notable that the Egyptian utilities tend to perform poorly, with five out of nine utilities performing near the bottom of the set. This suggests there may be some common factors behind their performance, such as an idiosyncrasy in compiling the data or a common national policy that leads to poor performance. (The Egyptian case study in part II of this book provides further insights.) This finding might not have been identified through the use of a single indicator and points to the value of using several dimensions of performance at the same time.

The spacing of the average rank values is also of interest. With 17 utilities, the maximum average rank score is 17 (one utility is best at everything) and the minimum is 1 (one utility is worst at everything). In practice, the values range from 12.4 to 4.4, and the gap in the average rank score between successive performers is about 0.5 points. EDCO is 0.8 points ahead of LYDEC, indicating clear superiority on the basis of the average range score criterion. At the other end of the distribution, AEDC is 1.8 points below the next-worst performer, indicating a markedly poor performance.

Finally, across the set, we find no indication that the utilities are trying to improve efficiency simultaneously on a subgroup of indicators. The extent to which utilities rank similarly on a particular indicator is measured by the coefficient of concordance. The average Spearman rank correlation between indicators is +0.008, so that the concordance is 0.20 and the probability of exceeding such a value (under the null hypothesis of no association between the different rank scores) is 0.42. This value far exceeds the conventional 5 percent used to reject the null hypothesis.

Generation Utilities: Average Rank Score

For GUs, the coverage of the indicators was much thinner. To include technical and financial indicators and a reasonable spread of utilities, we decided to retain only three indicators: the ratio of current assets to current liabilities, ROA, and the capacity factor. This set of indicators was available for 13 utilities.

Table 4.4 presents the individual and average rank scores for the GUs that provided data on all three indicators. The best performing is the Qatrana Electric Power Company (QEPCO) in Jordan, followed by the Al-Kamil Power Plant (AKPP) and the ACWA Power Barka (APBS), both in Oman. The worst performers are in Egypt: the Cairo Electricity Production Company (CEPC) and the West Delta Electricity Production Company (WDEPC). The score gap between QEPCO (12.0) and the next-best performer (9.7) indicates a very large difference in performance between these utilities and suggests that QEPCO is well in advance of the other GUs in the set. Three Egyptian utilities (out of the four for which there are data) occupy the bottom three places in the average ranking, suggesting that policy has not focused on improving performance even toward levels seen elsewhere in the MENA region.

Table 4.4 Ranks and Average Rank Score for Generation Utilities, MENA

Utility	Current assets/ current liabilities	Return on assets	Capacity factor[a]	Average rank
Jordan - QEPCO	13	10	13	12.0
Oman - AKPP	7	12	10	9.7
Oman - APBS	11	11	7	9.7
Oman - SPP	10	7	11	9.3
Oman - GPDCO	12	6	5	7.7
Jordan - CEGCO	8	13	1	7.3
Egypt, Arab Rep. - UEEPC	4	5	12	7.0
Egypt, Arab Rep. - MDEPC	6	4	9	6.3
Jordan - SEPCO	9	8	2	6.3
Oman - UPC	3	9	3	5.0
Egypt, Arab Rep. - EDEPC	2	2	8	4.0
Egypt, Arab Rep. - WDEPC	5	1	4	3.3
Egypt, Arab Rep. - CEPC	1	3	6	3.3

Source: World Bank calculations.
Note: MENA = Middle East and North Africa.
a. For an interconnected system.

The average Spearman rank correlation between the series is +0.21, which implies a concordance value of 0.47. The probability of exceeding this value, with 3 indicators and 13 observations, is 0.15. Therefore, the hypothesis of zero concordance between the indicators is accepted: utilities as a group show no tendency to perform well or badly across all dimensions of performance. They appear to focus randomly on certain indicators of performance and to pay less attention to other indicators.

Vertically Integrated Utilities: Average Rank Score

Four indicators covering all dimensions of performance (technical, commercial, and financial) were chosen for the average rank score: OPEX per connection, current assets to current liabilities, total energy sold per connection, and distribution losses. With this set of indicators, 8 of the total 12 VIUs in the sample could be included.

The results for the rankings and average rank score are shown in table 4.5. The best performance is that of the Saudi Electricity Company (SEC) in Saudi Arabia, followed by Algeria's Socièté Nationale de l'Electricité et du Gaz (SONELGAZ). The worst performance is that of Electricité du Liban (EdL) in Lebanon, followed by the Public Electricity Corporation (PEC) in the Republic of Yemen. The gap in the average rank score between the two worst performers is notably large (2–3.5 points), indicating that EdL's performance is particularly poor.

The average Spearman rank correlation is −0.06, implying a W value of 0.20. The probability of observing this value is 0.57, supporting the hypothesis that VIUs as a group did not focus on the performance of particular indicators.

Table 4.5 Ranks and Average Rank Score for Vertically Integrated Utilities, MENA

Utility	OPEX/connection	Current assets/ current liabilities	Total energy volume sold/connection	Distribution losses	Average rank score
Saudi Arabia—SEC	4	5	8	8	6.3
Algeria—SONELGAZ	7	8	5	3	5.8
Oman—RAECO	1	7	6	7	5.3
Tunisia—STEG	5	6	2	6	4.8
Morocco—ONEE	6	3	4	5	4.5
Oman—DPC	3	2	7	4	4.0
Yemen, Rep.—PEC	8	4	1	1	3.5
Lebanon—EdL	2	1	3	2	2.0

Source: World Bank calculations.
Note: MENA = Middle East and North Africa; OPEX = operating expenses.

Conclusion

The average rank score provides a method of identifying the better-performing utilities among a group that share a common set of data, and for which reliance on a single indicator could be misleading. In the case of MENA, the data gaps are substantial, which substantially reduces the number of utilities that could be compared. This effect was particularly notable for GUs: only 13 of the original 27 could be compared on a common basis.

For all three utility types analyzed, the coefficient of concordance between the series was low, and the null hypothesis of no agreement in rankings between series was accepted. This suggests that relying on the ranking of a single indicator would produce very different results than the same exercise using any other single indicator. Combining scores is more likely to provide a reliable picture.

Furthermore, low concordance values suggest that, generally, utilities were focusing on different subsets of indicators to improve performance. However, the clustering of poor performance scores for Egyptian DUs and GUs suggests that common policies are leading to poor performance within that country.

The average rank scores also identified utilities with extremely good or extremely poor performance by comparing them to the next-best (or worst) utility. QEPCO (Jordan) was well in front of the other GUs analyzed, EdL (Lebanon) was well behind other VIUs, and AEDC (Egypt) was far behind other DUs. The ability of the method to highlight such cases could inform subsequent analysis, by indicating which policies are factors of success and which of failure.

Notes

1. W can be calculated in alternative ways as shown in "Real Statistics Using Excel," http://www.real-statistics.com/reliability/kendalls-w/. Taking the average of the Spearman rank correlations (the usual correlation formula applied to the ranked values) of all pairs of indicator variables, denoted by r, m as the number of indicators, and k as the number of observations, then $W = \dfrac{(m-1) \times r + 1}{m}$. It can be shown that when there is complete agreement between indicators (the ranking is the same for every indicator), then W reaches its maximum value of unity. When there is no agreement between indicators—differences in rank scores between indicators are large—the minimum value of W is zero.

2. Under the conditions $m > 15$ or $k \geq 5$: $m \times (k - 1) \times W \sim \chi^2(k - 1)$ when the null hypothesis of no agreement is true.

3. Most indicators show better performance as they increase and are ranked in ascending order (1 = worst, 17 = best), but for indicators such as OPEX/km, where smaller values are better, the ranking is in descending order.

4. OPEX/km and revenue per employee are technical indicators, energy volume per connection is a commercial indicator, and energy sales per OPEX and return on equity are financial indicators.

CHAPTER 5

Drivers of Utility Performance: Institutional and Contextual Characteristics

The tremendous global heterogeneity of electricity sector structures may be one of the most striking stylized facts characterizing this sector.[1] The indicators collected for this study show that the Middle East and North Africa (MENA) region is no exception. Even if utilities continue to be central to each of the organizational models adopted in the region, these models differ across a number of institutional and contextual characteristics. Some of these differences have been credited with, or blamed for, differences in utilities' performance. Policy choices, such as the unbundling of the sector, the introduction of private ownership, or the introduction of a separate regulatory authority, have been suggested as key steps in improving the overall performance of the electricity sector (Bacon and Besant-Jones 2001). However, the lack of overwhelming evidence for the benefits of power sector reform as a panacea for poorly performing power utilities is leading to a reevaluation of policy responses to this underperformance.[2] At the same time, further evidence on the impact of various sector reform strategies can help inform the debate. The data collected for this analysis of sector performance in the MENA region provide the opportunity to contribute to this discussion.

The specific *institutional dimensions* related to the data on performance indicators collected are as follows: (a) the degree of vertical integration and the specialization of the utility (that is, the type of utility), (b) the size of the utility, (c) the nature of its primary ownership, and (d) the presence (or not) of a separate regulatory agency. One *contextual dimension* characterizing the environment in which the utility operates is added to this list, namely (e) the economy's overall level of income. Given that income levels have a highly negative correlation to energy imports in the economies of our study,[3] the correlation between performance of utilities and income level should be similar to the one between performance of utilities and an economy's net energy imports. This chapter relates these five dimensions to each of the performance indicators included in this study, in a first attempt to test for any connections (for example, to see if public and private utilities performed differently on a particular indicator).

Table 5.1 lists the set of utilities in the MENA Electricity Database, categorized by the institutional and contextual dimensions used for this analysis. The study divides utilities into four classes: vertically integrated utilities (VIUs), generation utilities (GUs), transmission utilities (TUs), and distribution utilities (DUs). The number of VIUs (12) is much smaller than the number of GUs (23) or DUs (29), and there are but a handful of TUs (3). The majority of DUs are found in the Arab Republic of Egypt (9) and Morocco (11), and the majority of GUs in Oman (12) and Egypt (5).

Table 5.1 Breakdown of Sample Utilities by Size, Ownership, Presence of a Separate Regulator, and Income, MENA, 2013 (or most recent year with data, 2009–12)

Categories	Measure	Vertically integrated utility	Distribution utility	Generation utility	Transmission utility
Size	Big	5	8	8	3
	Medium	4	10	6	0
	Small	3	11	9	0
Ownership	Public	11	21	10	3
	Private	1	8	13	0
Presence of separate regulatory agency	Present	6	18	23	3
	Absent	6	11	0	0
Income	High	5	3	12	1
	Upper-middle	4	3	6	1
	Lower-middle	3	23	5	1

Source: World Bank calculations.
Note: MENA = Middle East and North Africa.

Table 5.1 shows that the sample is relatively well distributed across sizes, because it includes 24 big, 20 medium-sized, and 23 small utilities. Some biases are more peculiar, such as the fact that there are no large private utilities in our sample: this points to a major difference between the MENA region and other regions of the world. The most obvious economy-related bias is that 14 of the 24 big utilities are Egyptian.

The table also shows that with respect to ownership, the sample is not well balanced. All TUs are public, so we cannot assess the impact of ownership. Similarly, all the big utilities are public because the big utilities include VIUs and TUs as well as other specialized utilities. However, the many DUs and GUs spread throughout the region have both public and private ownership.

There are 50 utilities operating in economies with a sector regulator, leaving 17 utilities without a sector regulator. However, none of the GUs (or TUs) are subject to a regulator, limiting the evidence of any impact on these utility types.

The high-income country (HIC) group includes 21 utilities, the upper-middle-income country (UMIC) group includes 14 utilities, and the lower-middle-income country (LMIC) group includes 32 utilities. Although the sample sizes of TUs and VIUs are small, they are spread evenly across income levels. By contrast, GUs are heavily concentrated in HICs, and DUs utilities in LMICs.

Potential Determinants of Utility Performance

Type of utility, organizational structure, and performance. The literature has emphasized unbundling vertically integrated power utilities as one step toward improving performance, and the horizontal unbundling of generation and distribution as a further performance-enhancing step. Unbundling generation from transmission and distribution (T&D) allows for multiple GUs, financed by private capital, and the introduction of some form of competition. These steps are expected to improve performance by reducing costs and increasing efficiency. Similarly, unbundling distribution allows the introduction of multiple utilities and private ownership and the possibility of competition, which again are expected to improve performance.[4] However, for small utilities, vertical and horizontal separation and the introduction of multiple entities reduces the average utility size and may result in the loss of economies of scale and scope. In the MENA region, although several economies have experienced unbundling, none has yet introduced competition between generators or between distributors. Hence any structural changes could not rely on gains from competition. Indeed, if scale economies are important, as is likely true for generation and transmission elements, then reducing the size of the average power company in an economy might be expected to worsen performance. However, there is a counterbalancing factor made possible by unbundling the functionally different components of a VIU. Managers with limited experience may find it simpler to concentrate on the key functions of generation or of distribution rather than to balance the conflicting interests of a VIU. In this case performance could be higher where unbundling has been introduced. Structure may then have a positive relation to performance even in the absence of competition or of private ownership. It is to be expected that this effect is weaker before the introduction of private capital prompts an intensified search for higher profits and lower costs and weaker still than it would be amid competition between utilities.

Size and performance. In a sector in which the existence of economies of scale and scope has been the working assumption, any significant performance differences according to size deserve a close look.

Size is notably varied across subsectors and contexts. Among other things, differences reflect policy makers' efforts to address climate change concerns and attract private investors to finance at least part of a utility's investment requirements, and eventually to introduce competition. These sources of pressure have had a significant impact on the way optimal structure is being discussed in the literature. Indeed, many observers argue for unbundling the sector to make the most of the latest renewable technologies, which would impact the optimal size of T&D. In a nutshell, the case for fragmenting the sector into smaller units seems to be growing. Therefore, we need to understand the extent to which current preferences leave room for improvement—and on which dimensions. Significant performance differences, especially in terms of costs, may explain some of the reluctance to restructure, as long as the new technologies or market structures cannot guarantee improvement. This chapter contributes to the discussion by clarifying differences in performance according to size. It provides a baseline on which to anchor the growing case for a redesign of the sector so as to make the most of renewable resources.

In summary, horizontal unbundling implies a reduction in the size of GUs and DUs. Until competition between utilities of the same type is introduced, the loss of scale due to unbundling may actually lower performance.[5]

One challenge in trying to assess the relevance of size is its subjective nature. In the context of this study, the following definitions have been adopted. The sizes of VIUs and DUs are defined by number of connections. A utility with fewer than 250,000 connections is small; utilities between 250,001 and 2 million are medium; utilities above 2 million are big. Because GUs do not have direct customers, the total installed capacity of power plants is the key determinant of utility size. For this purpose, GUs with installed capacity below 500 megawatts (MW) are considered small, those with installed capacity greater than 1 gigawatt (GW) are big, and anything between is medium. Finally, for TUs, the amount of energy transmitted determines size. Those transmitting less than 5 terawatt-hours (TWh) are small; those transmitting between 5 TWh and 10 TWh are medium; and those transmitting more than 10 TWh are big.

Ownership and performance. For almost 30 years, the debate on the relative effectiveness of the public and private operation of electricity utilities has been raging. It has yet to be settled. The experiences have been so diverse that there is no possibility of a definitive answer. Arguments for the benefits of private participation and ownership stress the pressure from new owners to maximize profits through efficiency and pricing strategies. Where prices are controlled, as in the MENA region, one expected effect is a reduction of costs.

A further benefit of allowing private ownership into the sector is that it provides a source of finance and thus lightens the government's financial burden, which is sure to grow heavier as demand for power increases. Also, the discipline of market financing is more likely to avoid the adoption of suboptimal projects. Governments may support projects for political rather than economic reasons, without paying attention to the costs of doing so. A further argument for encouraging the entry of private sector investment is that it sets an example of good management that publicly owned utilities may be encouraged to emulate.

Other aspects of performance may become secondary, provided that they are not seen as interfering with the return on investment. This chapter summarizes some basic, stylized results of an exercise in comparing the performance of electricity utilities in MENA depending on their ownership (that is, public or private). The discussion focuses on correlations rather than causality. In most cases represented in the MENA Electricity Database—if not all—a self-selection bias prevails and explains why some economies have gone one way and some another in terms of ownership.

Presence of a separate regulatory agency and performance. Creating independent regulatory institutions has been a standard component of electricity sector reforms for almost 20 years (see, for example, Jasmab and others 2015). Some MENA economies have jumped on the bandwagon, others not.[6] According to Cambini and Franzi (2013), in a review of the regulatory governance of Mediterranean economies—including many in the MENA region—this decision has mattered to the implementation of key policy decisions. Their research emphasizes the impact of separate regulatory agencies on the ability of MENA economies to attract investment to diversify energy sources. Cambini and Franzi do not, however, examine the impact on other more technical and specific performance indicators at the utility level and focus instead on the impact at the economy level.

This chapter provides additional insights on the impact of the decision to restructure regulatory governance in the MENA region by comparing (a) the performance of MENA electricity utilities supervised by separate regulatory agencies with (b) the performance of utilities operating in economies where regulation is still under the control of the sector ministry. To establish a possible link, we assess the correlation between any difference in performance across comparable utilities in the MENA region and the choice of regulatory governance at the very broad level, as allowed by the limited data on the detailed nature of this governance in the region. It is a weak test that does not establish causality between institutions and performance, but it manages to produce MENA-specific information in a region in which little related data have been collected.

The focus is on the broad signal offered by the institutional unbundling of the regulatory responsibility at the utility level. It does not get into the quality of the signal. As highlighted by Cambini and Franzi (2013) for their sample, the specific design of an institution may have a significant effect on the strength of the signal sent by its creation. The data collected here do not allow the internalization of these important dimensions. For instance, we do not consider the extent to which the separate regulatory agencies are financially or politically autonomous or the relevance of staff skills or the menu of mandates assigned to regulators and matching regulatory instruments. Despite this limitation, the research proves useful in assessing the extent to which a simple increase in the transparency of the regulatory function, allowed by the creation of a separate institution, made some difference—no matter what the quality of this institution was and how much of a difference it made. Indeed, the chapter shows that the impact of differences in regulatory governance is not binary, let alone simple (but this was to be expected, based on earlier assessments of international experience).

This work suggests that the impact of the introduction of a separate regulatory agency is difficult to anticipate. It varies significantly across economies and regions. In other words, context matters. In a survey of institutional reforms in the energy sector that includes a detailed assessment of the importance of regulatory institutions, complemented by additional analytical evidence, Vagliasindi and Besant-Jones (2013) show that the impact of introducing an independent regulator depends on a wide range of factors, including system size, development level, and demand composition. It also depends on the performance indicators being analyzed. For instance, an independent regulator may send a strong signal to investors without doing much to affect actual investment levels, and it may either increase or decrease prices, already refining some of the insights of Cambini and Franzi (2013).

Economy income level and performance. The performance of utilities may relate to the income level of the economy. For instance, demand for energy increases with income per capita, and this may change the composition of the demand base of the utilities. Growth usually comes with a stronger industrial sector, which tends to be more energy intensive. We also know that, in general, higher income levels are correlated with stronger institutions. This, in turn, may have an impact on the incentives utilities have to make stronger efforts to perform (that is, by reducing the risk of moral hazard in the management of the sector and among its various actors). It may also lead to access to more-experienced and better-equipped utilities (that is, reducing the risk of adverse selection by increasing the scope for competition in the sector, which is often associated with more cost-effective technical solutions).

To inform the discussion of this possible evolution, the sample has been divided into three groups: HICs, UMICs, and LMICs.[7]

Evidence of significant differences in performance according to income level indicate that simple comparisons of performance across economies, without taking into account differences in their income level, may be misleading. There may be other contextual factors that correlate with differences in utility performance and have not been explored in the context of this study on MENA utilities.

Summary of Results

Studies of the impacts of power sector reform have concentrated on a time-series approach—that is, for a particular economy or utility, the performance according to a number of indicators is compared prior to the introduction of the reform and for a number of years post reform. Jones, Tandon, and Vogelsang (1990) developed a method of comparing the historical and predicted future course of an industry with a counterfactual in which the industry remained unprivatized. Galal and others (1994) applied this method to two DUs in Chile, and Newbery and Pollitt (1997) described how to evaluate the restructuring and privatization of the U.K. electricity supply industry using this approach. In the latter case, great attention was paid to identifying those changes taking place that were due to external forces (for example, changes in European regulations) and those changes brought about by the act of privatization.

This approach focuses on one utility or one economy at a time and requires detailed knowledge of and data on the sector for a number of years before and after the policy change under evaluation. A crucial step in this type of analysis is the determination of how much performance would have changed in the absence of reform. Bacon and Besant-Jones (2001) quoted values for changes in performance on several indicators (energy sales, energy losses, employment, customers/employee, and net receivables) since privatization for four South American DUs. Treating such changes as entirely due to the effects of privatization is equivalent to assuming that without privatization there would have been no change in any of these indicators. For energy sales, certainly, this was an unrealistic assumption.

In the present study, the availability of data drawn from a large number of utilities exhibiting different characteristics provides the opportunity to test for the effects of various reform strategies in a different way. If the average performance of all public utilities on various indicators is poorer than that of the average for private utilities on the same indicators, then this supports the argument that privatization can help improve performance. In making such comparisons it is recognized that there are many individual factors that contribute to performance on a particular indicator, so that differences between public and private would not be due solely to their ownership status. A significant difference between performance levels across the two ownership types supports the argument that ownership matters. If the difference is not significant, this indicates that ownership does not in itself outweigh all the other factors determining performance on this particular indicator. But this does not prove that ownership has no impact on performance.

This chapter presents the results of an attempt to identify correlations, if any, between utility performance and five factors (type of utility, size, ownership, existence of a separate regulator, and income level of economy). A limitation of this exercise is that we only have cross-sectional and not time-series data, so no causality can be inferred. For each of the 36 performance indicators, the average for all relevant utilities over available observations is constructed. Next, for each of the five institutional and contextual factors, the averages for the same indicator are calculated and statistical tests of equality are carried out. For example, as a test of the importance of sector structure, data on the load factor from VIUs and DUs are tested to see whether the averages are the same for both utility types. Next, the mean load factor for utilities under a regulator is compared to the mean where there is no regulator, and so on. Results for the 36 performance indicators are grouped into the categories indicated in appendix A. Appendix D provides a brief account of the methodology used here.

For each indicator, table 5.2 specifies the classes of utilities to be included in the analysis, the total number of observations included in the test, and the overall mean for this indicator. It presents the probabilities of the tests for equality of means across five institutional and contextual factors (utility type, size, national income, ownership, and the presence of a seperate regulator). These are the probabilities of obtaining a difference at least as large as that observed if the null

Table 5.2 Tests of Equality between Subgroups of Factors Related to Indicator Mean Values (Probabilities) Using One-at-a-Time Testing, MENA Utilities

Classes of utilities included	Indicator	Category	Number	Mean	Utility type	Size	Income	Ownership	Separate regulatory agency present
VIU vs. DU	Load factor	System and operational efficiency	23	0.56	0.80	0.25	0.96	0.07*	0.63
VIU vs. GU	Capacity factor		20	0.54	0.07*	0.61	0.12	0.43	S
VIU vs. GU	Availability factor		11	0.93	0.50	0.71	0.04**	0.50	0.50
VIU vs. TU vs. DU	Network maintenance		10	0.02	0.79	0.41	0.85	0.52	0.20
VIU vs. DU	Share of meters replaced (%)		9	0.02	0.41	0.40	0.70	0.29	0.82
VIU vs. TU	Transmission losses	Losses efficiency	3	0.03	0.86	S	S	S	S
VIU vs. DU	Distribution losses		37	0.13	0.001**	0.76	0.52	0.63	0.69
VIU vs. DU	Technical losses		18	0.075	0.0003**	0.37	0.32	0.22	0.14
VIU vs. DU	Nontechnical losses		18	0.049	0.0003**	0.88	0.32	0.13	0.48
VIU vs. GU vs. TU vs. DU	OPEX/employee	Cost efficiency	48	274,000	n.a.	0.0001**	0.003**	0.006**	0.39
VIU vs. DU	OPEX/connection		36	723	n.a.	0.16	0.0001**	0.99	0.80
VIU vs. DU	OPEX/kWh sold		36	0.11	n.a.	0.002**	0.51	0.41	0.0001**
VIU vs. TU vs. DU	OPEX/km		37	24,381.0	n.a.	0.006**	0.95	0.02**	0.001**
VIU vs. DU	Residential connections/employee	Labor efficiency	24	238	n.a.	0.09*	0.54	0.15	0.02**
VIU vs. DU	Energy sales/employee		31	170,000	n.a.	0.03**	0.48	0.0007**	0.005**
VIU vs. DU	Total revenues/employee		34	212,000	n.a.	0.1*	0.70	0.004**	0.001**
VIU vs. GU	Cost fuels/OPEX	Cost structure	22	0.65	0.12	0.16	0.47	0.97	0.02**
VIU vs GU	Energy purchases + fuels/OPEX		8	0.77	S	0.05**	0.23	S	0.70
VIU vs. GU vs. DU	Labor cost/OPEX		35	0.13	0.22	0.03**	0.02**	0.13	0.29
VIU vs. DU	Energy sales/OPEX	Cost recovery	32	0.95	0.42	0.49	0.07*	0.83	0.15
VIU vs. DU	Energy sales/costs		19	0.82	0.11	0.13	0.03**	0.54	0.48

table continues next page

Table 5.2 Tests of Equality between Subgroups of Factors Related to Indicator Mean Values (Probabilities) Using One-at-a-Time Testing, MENA Utilities *(continued)*

Classes of utilities included	Indicator	Category	Number	Mean	Utility type	Size	Income	Ownership	Separate regulatory agency present
VIU vs. DU	Accounts receivable	Balance sheet	26	161	0.11	0.22	0.06*	0.84	0.63
VIU vs. GU vs. TU vs. DU	Debt/equity		47	7.08	0.24	0.05**	0.04**	0.62	0.67
VIU vs. GU vs. TU vs. DU	Assets/liabilities		53	1.17	0.32	0.0005**	0.31	0.56	0.84
VIU vs. GU vs. TU vs. DU	Return on assets	Profitability	49	0.3%	0.39	0.07*	0.22	0.05*	0.40
VIU vs. GU vs. TU vs. DU	Return on equity		46	4.6%	0.009**	0.10	0.15	0.03**	0.12
VIU vs. DU	Total energy volume/connection	Consumption and billing	35	6.4	0.002**	0.36	0.001**	98.0	0.21
VIU vs. DU	Residential energy volume/connection		23	4.0	0.01**	0.72	0.0001**	0.62	0.51
VIU vs. DU	Total billing/connection		27	297	0.17	0.005**	0.0001**	0.037**	0.09*
VIU vs. DU	Residential billing/connection		22	258	0.59	0.0001**	0.007**	0.37	0.34
VIU vs. DU	Collection rate		15	88%	0.03**	0.003**	0.86	0.51	0.08*
VIU vs. DU	Share of installed meters (%)	Metering	15	96%	0.32	0.33	0.02**	0.72	0.75
VIU vs. TU vs. DU	SAIFI	Customer management and service quality	15	1.6	0.02**	0.70	0.06*	0.37	0.69
VIU vs. TU vs. DU	SAIDI		12	28.6	0.46	0.35	0.72	0.49	0.57
VIU vs. TU vs. DU	CAIDI		9	52	0.21	0.46	S	S	0.20
VIU vs. TU vs. DU	Duration of interruptions		5	2.0	S	0.99	0.03**	0.32	0.03**

Source: World Bank calculations.

Note: Significant results are shaded in light red; performance indicators for which more than one factor gave significant results in one-at-a time testing are shaded in blue. CAIDI = Customer Average Interruption Duration Index; DU = distribution utility; GU = generation utility; km = kilometer; kWh = kilowatt-hour; MENA = Middle East and North Africa; n.a. = not applicable (tests are inappropriate); OPEX = operating expenses; S = singular dataset so estimation is not possible; SAIDI = System Average Interruption Duration Index; SAIFI = System Average Interruption Frequency Index; TU = transmission utility; VIU = vertically integrated utility.

Significance level: * = 10 percent, ** = 5 percent.

hypothesis of equality of the two means were correct. If the probability is less than 5 percent we concluded that there is a significant difference between the means.[8] Significant results are shaded in brown, and those tests that are inappropriate (for example, testing the effect of structure on operating expenses [OPEX]/employee) are shaded in blue. Those performance indicators for which more than one factor gave significant results in one-at-a-time testing are shaded in green.

Table 5.2 reveals that of the 36 performance indicators analyzed, as many as 25 have at least one factor that shows significant differences, and that there are 14 cases in which more than one factor is found to be significant in one-at-a-time testing. The substantial number of indicators for which there are significant results, even in the absence of detailed modeling of the situation, provides support for arguments that sector reform may be able to improve sector performance.

To focus on the relevance of these factors to performance, table 5.3 indicates which factors were most commonly related to performance, both by absolute number and as a percentage of indicators that could be tested for this effect. Three indicators (type of utility, size of utility, and the income level of the economy) were significant in around 30 percent of the cases, whereas ownership and the presence of a regulatory agency were significant in about 20 percent of the cases. Income was the factor found to be most often significantly related to performance indicators. In a region with wide variation in incomes from the LMIC to HIC levels, this serves as an important reminder that comparing performance across economies, without allowing for the effects of income level, could lead to a misjudgment as to the policy intervention required. Income is an important contextual factor: it has to be taken into account when designing sector reform policies, but does not point to any particular policy choice.

The four policy variables (utility type, size, ownership, and regulation) are significant often enough to suggest that based on the MENA utilities, with their wide range of individual circumstances, there is evidence to support the use of reform strategies that use vertical and horizontal unbundling, introduce private ownership, and create a regulatory body. The tests used to arrive at these conclusions support only broad approaches to policy—they do not distinguish, for example, between different types of regulatory bodies with different degrees of independence from the government. The contextual variable, income, is also

Table 5.3 Number of Indicators with a Significant Relation to Each Factor, MENA Utilities

	Type of utility	Size	Income	Ownership	Separate regulatory agency present
Number of significant results	8	11	12	6	7
% of significant results[a]	30	29	35	18	21

Source: World Bank calculations.
a. Percentage of significant results equals number of significant results relative to number of applicable indicators (that is total number minus not applicable and singular cases).

significant in several cases, making the point that the effects of policies may be dependent on the level of income in the economy concerned.

The results summarized in table 5.2 are from introducing one factor at a time into the tests for differences. A total of 14 indicators showed significant results for more than one driver. These cases were analyzed to see the effects of introducing more than one factor at the same time into the tests for differences. In 10 cases it was found that more than one factor is significant in testing for the simultaneous effects of several factors, and in 4 cases there was no support for the significance of more than one factor. These results suggest that a more detailed examination of performance—that is, introducing more contextual factors and refining the specification of the institutional factors—could provide further insights into the determinants of performance.

We then focus on indicator type. Grouping indicators into 11 categories and calculating the percentage of significant results by category yields the results in table 5.4. Certain categories of indicators show few significant links to the five factors, whereas others show a large number of significant links. For example, only 4 percent of the tests of system and operational efficiency are significant, compared with 56 percent for cost efficiency. These results suggest that the reform factors may be most often correlated with certain types of indicators.

Further insights are obtained by noting, for each driver, where there were significant results for a substantial proportion of the indicators within a given category. Table 5.5 shows that the significant results for each driver are concentrated within two or three indicator categories. For example, utility type has a substantial proportion of significant links to the losses efficiency, profitability, and consumption and billing categories, and no links at all to the systems and operational efficiency, cost structure, cost recovery, balance sheet, and metering categories. These results suggest that the effects of reform would not be felt across all indicators but are likely to be concentrated in certain aspects of performance.

Table 5.4 Number and Percentage of Significant Results, by Indicator Category

Indicator category	Number of indicators	Absolute number of significant results	Percentage of significant results
System and operational efficiency	5	1	4
Losses efficiency	4	3	14
Cost efficiency	4	9	56
Labor efficiency	3	6	50
Cost structure	3	3	23
Cost recovery	2	1	10
Balance sheet	3	3	20
Profitability	2	2	20
Consumption and billing	5	11	44
Metering	1	1	20
Customer management and service quality	4	3	18

Source: World Bank calculations.

Shedding Light on Electricity Utilities in the Middle East and North Africa
http://dx.doi.org/10.1596/978-1-4648-1182-1

Table 5.5 Categories of Indicators Whose Drivers of Performance Show Significant Results for a Substantial Proportion of the Indicators in that Category

Driver of performance	Categories with significant results
Type of utility	Losses efficiency, profitability, consumption and billing
Size	Cost efficiency, balance sheet, consumption and billing
Ownership	Cost efficiency, labor efficiency
Regulation	Cost efficiency, labor efficiency
Income	Cost efficiency, consumption and billing, metering

Source: World Bank calculations.

Statistically Significant Differences between Subgroups of Characteristics

Quasi-Fiscal Deficits and Drivers of Performance

The quasi-fiscal deficits (QFDs) described in chapter 1 measure performance through a combination of factors. This suggests that a general test of the relation between performance and the five drivers of performance can be made by relating total QFD, and each of its components, to the drivers. For this exercise, it is sensible to focus on utilities, and so the QFD (and each of its components) as a share of utility revenue is used. We have data on 9 VIUs[9] and 17 DUs. These groups are tested separately. (Further, the VIUs selected are all publicly owned so no test of ownership can be carried out for them.)

On the one hand, for VIUs, no performance driver is significant for the total QFD or for its components (underpricing, T&D losses, collection losses, and overstaffing). For the DUs, no test is significant for the total QFD, for T&D losses, and for collection.[10]

On the other hand, there are also some significant results for DUs. Regarding overstaffing, the big utilities had a significantly (1 percent probability) higher ratio of employees to total revenue (25 percent) than did medium (7 percent) or small utilities (8 percent). Utilities in LMICs had a significantly (probability 2 percent) higher share (21 percent) than in UMICs (2 percent) and in HICs (1 percent). Private utilities had a significantly (probability 3 percent) lower share (4 percent) than public utilities (19 percent), although regulation was not significant for overstaffing. Regarding underpricing, there is weak evidence (probability 10 percent) of a difference in the average share of revenue between private utilities (33 percent) and public utilities (64 percent).

The correlations between the QFD components and the drivers of performance provided clear evidence of links to overstaffing in DUs, suggesting that private ownership is associated with a smaller degree of overstaffing. The other components of the QFD were not correlated with the drivers. This cannot be taken as a rejection of the relevance of reform strategies for the power sector, but rather indicates that a more fully specified analysis of the links to performance would be needed before such a determination could be made.

Utility Type and Drivers of Performance

For some indicators, differences across utility types are expected. OPEX covers a wider range of functions for a VIU than for a DU serving the same number of customers, because it has to incur costs for generation and transmission activities. Hence it would be meaningless to test for equality of utility type for any indicator incorporating OPEX.

The broad picture of the impacts of reform indicate that utility type is significantly linked to losses efficiency, profitability, and consumption and billing indicators, and not to system and technical efficiency, cost structure, cost recovery, and balance sheet indicators.

Distribution losses averaged over all utilities at 13 percent, which is similar to the range of values found in the non-MENA group. However, the data show that DUs have a significantly better performance level (10 percent) than VIUs (20 percent). This suggests that DUs are better able to focus on their primary business than VIUs, whose problems are more widespread. This provides support for those arguing that unbundling can stimulate cost reductions.

Technical losses are made up of nonvariable technical losses and variable technical losses. These were significantly lower for DUs (7 percent) than for VIUs (10 percent). Technical losses may reflect a relatively low load factor, because consumption (and therefore load) is less even throughout the day.

Nontechnical losses were on average 4.9 percent, and VIUs (10 percent) showed much larger losses than DUs (3.6 percent), indicating a significant difference between the performance of these two groups.

Return on equity (ROE) stands at 4.6 percent. The highest ROE is observed for GUs (11 percent), followed by DUs (7.0 percent), and VIUs (−23.0 percent). The Omani TU's ROE stands at 20 percent, but this is not representative of all TUs in the region. Significance tests indicate that ROE for GUs and DUs is significantly higher than for VIUs and lower than for TUs. VIUs in the MENA region perform poorly in terms of ROE compared with those outside MENA. Risk perception would be high in any environment in which domestic tariffs need to be subsidized or depend on politically sensitive cross-subsidies, as is the case for most VIUs in MENA.

Return on assets (ROA) and ROE, on average, would not pass the common hurdle rates considered by investors and lenders. This is even true for the hurdle rates adopted by most international organizations, whether they want to support public or private projects. This exposes economies to increased risk of projects ending up being packaged to meet these hurdle rates rather than to address the broader investment challenges of the sector.

The average *total energy volume sold per connection* (and year) is generally quite reasonable by international standards, accounting for the income level and the industrial and service structure of the region. The VIUs (12.6 megawatt-hours [MWh]) sell a significantly greater amount per connection than the DUs (4.2 MWh). This probably reflects the nature of the customers served by these different types of utilities, rather than reflecting an inherent superiority of VIUs.

Shedding Light on Electricity Utilities in the Middle East and North Africa
http://dx.doi.org/10.1596/978-1-4648-1182-1

The *residential energy volume sold per connection* is also consistent with the international best practice in economies with similar income characteristics. The differences between VIUs (8.0 MWh) and DUs (2.6 MWh) are again significant.

The *collection rate* is the ratio of the revenue collected to the total electricity billed. The higher the ratio, the higher the effectiveness of the utility in bill collection. DUs have a significantly higher collection rate (89 percent) than VIUs (69 percent), probably because their narrower focus frees them to focus more on collection.

The average *System Average Interruption Frequency Index (SAIFI)* illustrates, at best, a reasonable performance. The SAIFI for DUs (1.23) is significantly lower than that for VIUs (3.18).

This group of indicators, for which there are significant differences in performance, offers a coherent picture: utilities that have only a distribution function are able to concentrate on reducing losses and improving collection and perform better than VIUs with respect to these indicators.

Size and Drivers of Performance

Size was found to be significantly linked to cost efficiency, balance sheet, and consumption and billing indicators. It was not linked to system and operational efficiency, losses efficiency, cost recovery, profitability, and customer management and service quality indicators—these are areas where the degree of government support is unlikely to impact performance but where management quality can have a significant effect.

OPEX per employee differs quite significantly by type of utility. Values vary, from $376,000 for GUs, $190,000 for DUs, $216,000 for VIUs, to $58,000 for TUs. This result is in line with what practitioners expect. The hypothesis that OPEX per employee is constant across sizes is rejected. For the MENA region, the value for this indicator is significantly smaller for big utilities ($99,000) than for medium ($410,000) or small ($236,000) utilities, although the latter two were not significantly different. This result supports the notion of economies of scale being important at higher levels of operation. This needs to be put in context. The average number of employees is almost seven times higher among big utilities. Some big utilities, such as Société Nationale de l'Electricité et du Gaz in Algeria, employ almost 20,000 people, whereas small utilities in Djibouti, for example, have about 1,000 employees.

When comparing utilities in the MENA region by size, large utilities have a significantly lower *OPEX per kilowatt-hour (kWh)* ($0.04) than medium ($0.13) or small ($0.15) utilities. This is consistent with the suggestion that economies of scale are still strong in some economies of the region.

In comparisons of *OPEX per kilometer (km)* by size of utility, the value for big utilities ($12,753/km) was significantly lower than for medium ($35,179/km) and small ($27,896/km) utilities.

Size also shows a significant difference for the share of **labor cost in total OPEX**, between the big (17 percent), medium (10 percent), and small (12 percent) utilities.

A test of the relation between the **debt-to-equity ratio** and the size of the utility indicated that the ratio for big utilities (1,143 percent) was significantly higher than for medium (499 percent) and small (330 percent) utilities. Even though they are the least leveraged in the region, small utilities are still very highly leveraged by international Organization of Economic Co-operation and Development (OECD) standards in which leverage is usually less than 100 percent. Even if the current low levels of interest should provide a good margin to rely on debt finance, given the risk premia and the long-term nature of the financial commitment often indexed to price changes, the current MENA approach appears to be risky. The prospects for high leverage are, however, probably better for smaller utilities if their cost-recovery performance continues to be solid. For the larger utilities, costly government financing or guarantees continue to be the main option to stay highly leveraged.

The ratio of **current assets to current liabilities** also differs according to size. Big utilities have a ratio of 79 percent, whereas medium are at 84 percent and small at 200 percent; and the difference between the large and medium and the small subgroups is statistically significant. In the MENA region, the larger the utility the less likely it is to be able to pay off its short-term liabilities. This reinforces the conclusion that the MENA region's smaller utilities are better managed financially than the larger ones.

The average total **billing per connection** is lower for DUs ($268) than for VIUs ($392), but this difference is not statistically significant. Total billing per connection is significantly related to the size of the utilities. Big utilities ($155) have lower billing per connection than do medium ($404) or small ($381) utilities. Similar results are found for **residential billing per connection**.

The **collection rate** is the ratio of the revenue collected to the total electricity billed. The higher the ratio, the higher the effectiveness of the utility in bill collection. The collection rate is significantly related to size, with big utilities at 91 percent, medium at 96 percent, and small at 65 percent. This result is somewhat unexpected and may be due to the small sample size and problems in measuring this variable.

Size is significant for a group of indicators that include OPEX as a component. For **OPEX/employee, OPEX/kWh, and OPEX/km,** large utilities have the lowest value and are the most efficient. Where OPEX is in the denominator, as for labor costs/OPEX the large utilities have the highest value, indicating that OPEX rises slower than labor costs as utility size increases. However, care has to be taken in interpreting these results. One of the major components of OPEX is fuel costs and the pricing of fuel across the region is by no means uniform. Large utilities may be concentrated in countries where the largest energy subsidies are available. The test for a size effect on the ratio of fuel costs did not reveal significant differences, and further detailed analysis to understand these results could yield useful insights. This interlinking with government support is probably reflected in the relationship between size and the debt/equity and current assets/current liabilities ratios. Big utilities have the highest debt/equity ratios and the lowest current assets/current liabilities, both of which indicate a weak financial position, probably made possible by government support.

These significant results for the relation of size to performance do not add support to the existence of technical economies of scale but are very important in indicating how in MENA government policy may have supported the larger utilities allowing them to support more adverse financial performance.

Ownership and Drivers of Performance

The state of ownership of a utility—private or public—has been one of the areas of discussion with respect to improving utility performance. Private ownership introduces the profit motive and incentives for improving performance. Privatization alone is recognized to run the danger of creating private sector monopolies where profits are increased but at the expense of the consumer by allowing prices to rise while decreasing costs. Where competition can also be created then the dangers are less, but the complex market structures and institutions required to permit full competition do not exist within MENA, or in many other countries. Accordingly increasing reliance is placed on government control possibly through the creation of a regulatory body.

Ownership was found to be significantly correlated with cost efficiency, labor efficiency, and the ROE indicators. It was not correlated with system and operational efficiency, losses efficiency, cost structure, cost recovery, and balance sheet indicators.

Public utilities have significantly lower **OPEX per employee** ($180,000) than private utilities ($417,000). This is in a region in which utilities' OPEX are largely dominated by the costs of fuel and labor. Although OPEX of public utilities is on average three times larger than that of private utilities in the MENA region, private utilities in the sample have almost six times fewer staff than public utilities. This would more than offset the differential in OPEX between the two categories. This result strongly supports the view that private ownership can result in the reduction of costly overstaffing. **OPEX/km,** too, is significantly lower for public utilities ($20,166/km) than for private ones ($37,290/km).

There is a significant difference between **sales per employee** across private ($341,000) and public ($126,000) utilities. The public utilities are hampered by their much higher levels of employment. Private utilities had significantly higher **revenues per employee** ($360,000) than public utilities ($167,000). As is the case with sales per employee, this result supports the view that privatization does improve efficiency by reducing employment.

The **ROE** among publicly owned utilities (0 percent) is well below that of privately owned (15.5 percent), indicating that public utilities are able to withstand a low ROE because of government support. The magnitude of this difference suggests that the cost of this support must be substantial.

For **total billing per connection** there is a significant difference between public ($259) and private ($464) utilities, easily explained by differences in the absolute number of connections. Public utilities in MENA have one-and-a-half times the value of total sales than private utilities, but almost four times the number of connections.

It is notable that ownership type was not significant for any indicators of system and operational efficiency, losses efficiency, or customer management and service quality. These are activities where private management might have been expected to improve performance by introducing better procedures and more modern technology. The substantial improvements in private sector performance appear to come from small staff size. This probably reflects the use of state-owned enterprises (SOEs) as a means of absorbing some of the otherwise unemployed labor force.

Regulation and Performance

One of the main arguments for introducing a regulatory authority is to exercise some control over how VIUs and DUs set prices. This control, in turn, is expected to reduce costs as utilities look to maintain or increase profit margins. Improvements in system and operation efficiency, losses efficiency, and consumption and billing are expected under regulatory control.

Table 5.5 indicates that regulation is significantly correlated with cost efficiency and labor efficiency indicators, but not with system and technical efficiency, losses efficiency, cost recovery, balance sheet, profitability, and consumption and billing indicators.

Utilities operating where there is a separate regulatory agency have a significantly lower **OPEX/kWh** ($0.07) than utilities operating without one ($0.15). Similarly, for **OPEX/km**, utilities with a separate regulator have significantly lower values ($16,944/km) than those without ($36,469/km).

Utilities with a separate regulator have far fewer residential **connections per employee** (205) than those without such a regulator (472). It is implausible that regulation would lead to a reduction in connections per employee (and an increase in employees per connection), so this result is likely circumstantial. The same goes for **energy sales per employee:** utilities with a separate regulator present had sales of $117,000 and with no regulator had sales of $279,000 per employee. Also, for **total revenues per employee,** utilities operating with a separate regulator had lower revenue per employee ($132,000) than those with none ($327,000).

The average **share of cost of fuel, lubricant, gas, and coal in total OPEX** is surprisingly high in an oil- and gas-producing region, at 66 percent, and utilities operating with a separate regulatory agency have a significantly higher average share (70 percent) than utilities operating with none (45 percent).

The presence or absence of a regulatory agency is not correlated with any indicator of system and operational efficiency, losses efficiency, balance sheet, profitability, or consumption and billing. The correlations that do exist are probably circumstantial.

Income and Drivers of Performance

Income is not a direct policy instrument, but may help explain government attitudes toward the power sector. Richer countries are better able to provide financial support (directly or indirectly) and considerations of political economy within such countries may lead governments to provide such support. Lower tariffs meant to benefit consumers, or fuel input prices below international equivalents can play an important role in the performance of power utilities.

Table 5.5 indicated that income was significantly correlated with cost efficiency, consumption, and billing and metering indicators. It was not correlated with losses efficiency, labor efficiency, and profitability.

The *availability factor* is high, which suggests that the plants of the region can provide energy to the grid most of the time. However, although the availability factor is similar in HICs (92 percent) and in UMICs (98 percent), this difference is statistically significant (there are no observations on this indicator for LMICs). This suggests that in the HICs maintenance is less effective and the plants may be older (this conclusion may be influenced by the fact that most plants with low availability levels are in Oman, where weather conditions are extreme, with implications for peak load factors).

Regarding **OPEX/employee,** the LMICs spend the least ($159,000/employee), UMICs spend in the middle ($293,000), and HICs spend the most ($400,000), and the differences between these subgroups are significant. These differences could reflect employee performance levels or could be explained by other factors. For example, per capita labor and some other input costs increase with income level. **OPEX/connection** is also significantly related to income. The mean value for HICs was $1,993, whereas for UMICs it was $839, and for LMICs, $394.

Differences across income groups are also significant when considering the **share of labor costs in total OPEX**. LMICs have a share of 16 percent, UMICs have a share of 6 percent, whereas HICs have a share of 11 percent. The lower share seen in HICs might have structural reasons: four out of five HICs in the sample have a vertically integrated market.

There is significant difference across income levels in the **energy to sales ratio**, with the highest ratio in the LMICs (0.91) and the lowest ratio (0.56) in the HICs. Again, the government support offered to utilities in HICs is likely to explain this.

There was also a significant difference in the **debt equity ratio** between HICs (376 percent) and LMICs (1,065 percent). It appears that higher levels of economic development can lead to more acceptable levels of risk.

The income level does appear to have an important effect on **energy sales per connection** (as would be expected). The average for HICs

(28.8 MWh) is significantly greater than for UMICs (5.7 MWh) and LMICs (3.9 MWh). For instance, HICs have high billing per connection, generally related to high energy per capita. Air conditioning is significant, particularly in economies in the Gulf Cooperation Council such as Bahrain and Saudi Arabia. The energy billed by LMICs is twice that of HICs, and LMICs also have five times the number of connections. Many utilities operate between $100 and $300 per connection. Because at least 50 percent of utilities in LMICs have an average billing of $300 per connection, these sales data provide further evidence of the impact of price controls and subsidies in LMICs. LMICs also have the lowest average residential billing per connection ($168), yet residential users represent a very large market segment, particularly in economies where self-generation in the industrial sector continues to be a common solution. Residential sector sales account for a large percentage of billing in most economies in the MENA region.

The *residential energy volume per connection* shows a significant difference across HICs (17.9 MWh), UMICs (4.1 MWh), and LMICs (2.3 MWh). The reasons are the same as for total sales per connection. Income levels are also significant for *total billing per connection*—HICs ($925) have larger values than UMICs ($419) and LMICs ($245); and for *residential billing per connection* (HICs are at $478, UMICs at $313, and LMICs at $168).

The *percentage of installed meters* is important: to measure consumption and manage demand when required, metering of all consumption points is needed. A large majority of customers have a meter (96 percent). If this is representative across the region, it indicates strong performance by international standards. Notably, the percentage in HICs (57 percent) is significantly below that in UMICs and LMICs (100 percent); this probably reflects government policy toward consumers in some HICs in the sample.

Information on the *duration of interruptions* indicates that values are significantly higher in UMICs than in LMICs. However, the sample of utilities answering this question is extremely small, so little import should be attached to this result.

The results obtained suggest that income is likely to affect certain performance indicators in two ways. Higher-income countries may decide to support consumers by reducing tariffs via some form of subsidy. This then affects performance on indicators that relate to revenues or costs. Also, higher-income countries tend to have greater demand for power, often met by larger utilities. Where there are genuine scale effects in supply costs, then performance will tend to be better in higher-income countries.

The presence of a relation between income and performance in MENA indicates that care should be taken when comparing performance across utilities without taking national income levels into account, and also in trying to understand the nature of the relationship between economy income and utility performance.

Conclusion

The primary objective of chapter 5 was to use evidence on utility performance from the MENA Electricity Database to explore whether cross-sectional (inter-utility) differences in performance are correlated with key institutional and contextual variables. A number of important conclusions can be drawn from the results described above:

1. The tests carried out, despite lack of data for certain indicators, reveal a substantial number of cases where performance indicators are correlated with one (or more) of the drivers—25 of the 36 indicators had some significant link with a driver of performance, and 14 of these had significant correlations with more than one driver.
2. About 30 percent of indicators had significant correlations to utility type, utility size, and national income; about 20 percent had the same to ownership (public or private) and the presence of a separate regulator.
3. Approximately 50 percent of results across three categories of indicators (cost efficiency, labor efficiency, and consumption and billing) were significant; the same was true for only 4 percent of results for system and loss efficiency.

These results provide evidence of links between drivers and performance overall, without taking individual circumstances into consideration. The bunching of significant results by indicator category suggests that certain areas of utility performance are more affected by policies linked to utility type, size, ownership, and regulation, whereas other areas of performance show few links to these policy-related drivers.

Organizational Structure

Unbundling the power sector has been said to permit more focused management and to increase the possibilities of competitive behavior once a market is liberalized. Utility type was significant for only a few indicators,[11] but for these the results were coherent and highly significant. DUs performed much better than VIUs in distribution losses (10 percent versus 20 percent, respectively) and in technical and nontechnical losses, analyzed separately. The differences were also significant for collection rate (89 percent versus 69 percent, respectively) and for SAIFI (1.23 versus 3.18, respectively). The ROE for DUs was 7.0 percent, whereas that for VIUs was −23 percent. Energy sales per connection were significantly higher for VIUs (12.6 MWh) than for DUs (4.2 MWh), and there was a similar result for residential sales per connection. Testing for size differences revealed that size was not significant for these variables, so that the differences between VIUs and DUs could not simply be assigned to scale. More likely, DUs were established in areas where sales per connection tended to be lower (especially for nonresidential customers).

The results suggest that DUs were better able to focus on business with end consumers and thus able to focus efficiency drives in a meaningful fashion. VIUs are more broadly focused, and their role as a national provider means that they may be required to pursue goals such as increasing employment, keeping consumer tariffs low through cross-subsidies, and keeping nonpaying customers connected.

Size

In considering power sector reform strategies, options such as unbundling VIUs and introducing more than one utility of the same type to better focus on core business and introduce a form of competition will reduce average utility size. Traditional analysis of the power sector has emphasized the role of economies of scale when discussing long-run pricing strategies. The factor of size can pull in two directions, so there is interest in considering the importance of size in the MENA context.

The size factor was significant across about one-third of indicators, and it was notably insignificant for system and operational efficiency and losses efficiency indicators. Values for OPEX/employee, OPEX/kWh, and OPEX/km indicate the potential importance of scale economies. The group of big utilities had values significantly lower than the medium and small utilities (and the differences between these two groups were not significant). A similar pattern was found for energy sales per employee and total revenues per employee.

The debt-to-equity ratio was much higher for big (1143 percent) than for medium (499 percent) and small utilities (330 percent), whereas the big utilities had a lower assets-to-liabilities ratio (79 percent) compared with medium utilities (84 percent) and small utilities (200 percent). Total billing per connection was ($155) for big utilities, whereas the values for medium ($404) and small ($381) utilities were significantly higher. These results suggest that size relates to performance in two ways. Economies of scale lead to lower OPEX per normalized scale factor (employee number, kilowatt-hour, kilometer), but that larger utilities also maintain very high debt-to-equity ratios, low assets-to-liabilities ratios, and low billing rates per connection, probably through the use of government finance and guarantees.

Ownership

Introducing private ownership is one of the main policy tools for improving the performance of power utilities. The direct introduction of the profit motive can be expected to prompt utilities to reduce costs as well as increase sales. One of the most direct ways to reduce costs may be to tackle overstaffing. Although public and private performance differed on only a few indicators, it was clear that private plants utilize less labor to achieve the same production.

OPEX/employee was significantly higher where there was private ownership ($417,000) as opposed to public ownership ($18,000). OPEX itself was three times higher for the public utilities, but employment was six times as high—suggesting considerable overstaffing. Similarly, significant differences were found for OPEX/km, total revenues per employee, and sales per employee. Total billing per connection was significantly higher for private ($464) than for public ($259) utilities.

The ROE was also significantly higher for private (15.5 percent) than for public (0 percent) utilities, which is consistent with private utilities placing more emphasis on the profit motive. A major difficulty with the use of this indicator in cross-section studies is the selection bias. Governments do not privatize a random selection of some or all of the utilities in the sector. Rather, they may select those that are already well performing, on the basis that these will be easier to privatize and are the ones in least need of continuing government support. This selection pattern produces a positive correlation between performance and ownership—but, notably, it is not a causal link.

Presence of a Separate Regulatory Agency

How the presence of a separate regulatory agency affects utility performance depends on what is being regulated. If the primary focus is the tariff level, then governments that wish to set tariffs low for political reasons, through the use of subsidies, will not introduce a regulator. This produces a positive correlation between the presence of a separate regulator and certain indicators (for example, energy sales to OPEX). Thus, the presence of a separate regulator has to be analyzed in context.

A group of indicators was found to have significantly lower values for utilities operating in the presence of a separate regulator than for those without one. These indicators included OPEX/kWh ($0.07 versus $0.15, respectively), OPEX/km ($16,944/km versus $36,469/km, respectively), residential connections per employee (205 versus 472, respectively), sales per employee ($117,000 versus $279,000, respectively), and revenues per employee ($132,000 versus $327,000, respectively). There is no apparent reason why the presence of a separate regulatory agency should produce such differences, and it is more likely that context is the deciding factor.

Income

The income level of the economy in which a utility is situated may well influence the utility's performance independent of size and structure. Two effects may be involved: (a) at higher per capita incomes, the consumption of electricity per household increases steadily, with an income elasticity of around unity and (b) in economies with high income levels, governments may be more willing to subsidize utilities to keep consumer tariffs low.

In the present study, the income factor was significant for one-third of the indicators. OPEX per employee increased significantly with income: the value for LMICs ($159,000) was significantly lower than for UMICs ($293,000) and HICs ($400,000), and OPEX per connection showed a similar pattern. Labor costs composed a larger share of OPEX in LMICs (16 percent) than in HICs (11 percent). An increase in OPEX per connection as an economy's income level rises to that of an HIC is to be expected: much of the demand increase as incomes rise is due to the same households purchasing more, rather than to low-income households deciding to connect to the grid (thereby increasing the number of connections but decreasing the average consumption per connection). The fact that energy volumes sold per connection increase sharply with income (HICs at 28.8 MWh and LMICs at 3.9 MWh) is more evidence of the impact of income on demand. A rise in OPEX/employee is consistent with the existence of higher wages per employee at higher income levels, and labor costs as a share of OPEX could fall for the same reason.

Energy sales as a percentage of costs declined from 91 percent in LMICs to 56 percent in HICs, probably because of the willingness of HICs in the region to charge lower tariffs and subsidize the utilities. The ratio of debt to equity was significantly higher in LMICs (1,065 percent) than in HICs (376 percent), suggesting that more-developed economies were able to work with more acceptable levels of risk.

Notes

1. See Jamasb, Nepal, and Timilsina (2015) for a broader review and Vagliasindi and Besant-Jones (2013) for a detailed analysis of organizational structures in low-income countries in the power sector.

2. Vagliasindi and Besant-Jones (2013) show that unbundling can deliver performance improvements, but not for all indicators. They emphasize that unbundling works best when part of broader reforms (for example, regulatory reforms and increased competition in generation and distribution) and for large systems in countries with a certain threshold of development (as measured by per capita income). They also find that partial unbundling is not effective.

3. Using 2013 World Development Indicators data, the correlation between net energy imports (as a percentage of energy use) and gross domestic product per capita (in purchasing power parity) is −0.77, which indicates a high correlation between these two variables.

4. Bacon and Besant-Jones (2001) describe the traditional approach to power sector reform, whereas Eberhard and Gratwick (2011) discuss how this approach has since evolved. Vagliasindi and Besant-Jones (2013) provide a more recent evaluation of approaches to power sector reform.

5. The need to tailor reform strategy to the size of the power system was argued by Bacon (1995).

6. Eight economies out of the 14 considered in this study have an electricity regulator, whether it is independent or not.

7. HICs are those with a gross national income (GNI) per capita above $12,736; UMICs have a GNI between $4,126 and $12,735; and LMICs, a GNI in between $1,046 and $4,125 (see World Bank Country and Lending Groups Database at https://datahelpdesk.worldbank.org/knowledgebase/articles/906519-world-bank -country-and-lending-groups; and World Bank Open Data at http://data .worldbank.org).

8. A value greater than 5 percent indicates that there is no difference between the means. Probability values of 10 percent or less are noted as indicating weak support for significant differences between the subgroups.

9. Excluding Iraq, for which the values were so large as to dominate any common relationships.

10. A weak effect—showing private distributors with a higher ratio of collection losses to total revenues than public utilities—was due to a single observation on the West Bank's Tubas distributor, where collection losses were 144 percent of total revenue. No significant difference was found when this observation was omitted.

11. It should be noted that testing for structural differences was not meaningful for seven indicators.

References

Bacon, R. 1995. "Appropriate Restructuring Strategies for the Power Generation Sector: The Case of Small Systems." Industry and Energy Department Occasional Paper 3, World Bank, Washington, DC.

Bacon, R., and J. Besant-Jones. 2001. "Global Electric Power Reform, Privatization and Liberalization of the Electric Power Industry in Developing Countries." *Annual Review of Energy and Environment* 26 (1): 331–59. Also as: Energy and Mining Sector Board Discussion Paper 2, World Bank, Washington, DC.

Cambini, C., and D. Franzi. 2013. "Independent Regulatory Agencies and Rules Harmonization for the Electricity Sector and Renewables in the Mediterranean Region." *Energy Policy* 60 (September): 179–91.

Eberhard, A., and K. N. Gratwick. 2011. "IPPs in Sub-Saharan Africa: Determinants of Success." *Energy Policy* 39: 5541–49.

Galal, A., L. Jones, P. Tandon, and I. Vogelsang. 1994. *Welfare Consequences of Selling Public Enterprises: An Empirical Analysis*. New York: Oxford University Press.

Jamasb, T., R. Nepal, G. R. Timilsina. 2015. "A Quarter Century Effort Yet to Come of Age: A Survey of Power Sector Reforms in Developing Countries." Policy Research Working Paper WPS 7330, World Bank Group, Washington, DC.

Jones, L., P. Tandon, and I. Vogelsang. 1990. *Selling Public Enterprises: A Cost-Benefit Methodology*. Cambridge, MA: MIT Press.

Newbery, D., and M. G. Pollitt.1997. "The Restructuring and Privatization of the U.K. Electricity Supply—Was It Worth It?" Public Policy for the Private Sector, Note 124, World Bank, Washington, DC.

Vagliasindi, M., and J. Besant-Jones. 2013. *Power Market Structure: Revisiting Policy Options. Directions in Development*. Washington, DC: World Bank.

PART II

What Do the Country Case Studies Tell Us?

These four case studies (of the Arab Republic of Egypt, Jordan, Morocco, and Oman) offer insights relevant to the Middle East and North Africa (MENA) region and beyond. The studies aim at providing not only an overview of each country's power sector but also an analysis of utility performance to help identify potential areas of improvement. The narrative and figures presented in these chapters focus on the year 2013 (as did part I). Although previous chapters compared utilities regionwide and across a range of performance indicators, chapters 6 to 9 compare utilities with one another and also with regional median values.

The four countries chosen for the case studies have undertaken significant reforms of their electricity sectors over the past decades. These countries have a wide variety of characteristics and challenges representative of the 14 MENA economies of this study. In a region where the sector is mostly publicly owned and centralized under vertically integrated utilities (VIUs), Egypt, Jordan, Morocco, and Oman each have a story to tell, whether in relation to their dependence on fossil fuel imports, their population size and geographical spread, or the initial and organizational structure of their electricity sector. As illustrated in appendix B (table B.1), Egypt, Jordan, Morocco, and Oman had gone through some degree of unbundling in their electricity sectors in 2013. By then, private sector involvement was well developed in Jordan and Oman. Egyptian utilities remained state owned with the exception of some independent power producers (IPPs) for generation. Morocco's electricity sector structure involved a single VIU (Office National de l'Electricité et de l'Eau Potable, ONEE) with the electricity distribution activities of most cities being delegated to 11 municipal entities, of which four are privately owned.

A number of factors exogenous to the electricity sector have affected the performance of the region's utilities. These factors include—but are not limited to—political instability (as in Egypt over the years 2011–14), disruptions in primary fuel supply (as in Jordan, where the entire sector was reformed due to gas supply interruptions, resulting in a radical shift in the energy mix), and both the direct and indirect spillover effects of regional armed conflicts (an influx of displaced populations from the Levant in countries such as Jordan, for instance, have resulted in a stark increase in population and, consequently, demand).

The case studies cover countries that have addressed, in different manners, the link between water and energy, which cannot be left unmentioned in the MENA region. Desalination plants are an integral part of the energy sector in the member countries of the Gulf Cooperation Council (GCC), supplying both municipalities and industries for the past two to three decades (Al Hashemi and others 2014). Also, several energy utilities are involved in water or sanitation activities. These two trends can be observed in Oman, where desalination activities are common among several electricity generation utilities (GUs), and in Morocco, where the 11 distribution utilities (DUs) are also involved in water and sanitation activities.

Also of interest is the introduction of renewable energies in the energy mix, in a region in which fossil fuels remain the dominant source of electricity, mostly due to their abundance and the conventional generation technologies and practices that have been in place for several decades. In 2013 in Morocco, 31 percent of total installed capacity was from renewables (of which 7 percent was not hydropower). Oman, in contrast, depended entirely on thermal power generation, with natural gas and diesel oil making up 98 percent and 2 percent of the energy mix, respectively (AER 2014). With several members of the MENA region benefiting from an abundance of solar and wind resources, the region's potential has yet to be exploited and is lagging behind other world regions mostly because renewable energy sources are disregarded in policy design.

Each of the four case studies here start out with a brief historical overview before detailing the main characteristics of the electricity sector's three main activities: generation, transmission, and distribution. This overview is followed by a discussion of the relative performance of GUs and a discussion of DUs. The scope is limited by the availability of data. Yet this represents a good start at developing analysis that might, in turn, inform ways to address the major challenges identified in this report. A synthesis of the evolution of the sector from 2014 until the writing of this book in 2017, which in some cases has gone through important reforms, is briefly presented before the concluding section.

References

AER (Authority for Electricity Regulation). 2014. *Annual Report 2014*. Muscat, Oman: AER.

Al Hashemi, R., S. Zarreen, A. Al Raisi, F. A. Al Marzooqi, and S. W. Hasan. 2014. "A Review of Desalination Trends in the Gulf Cooperation Council Countries." *International Interdisciplinary Journal of Scientific Research* 1 (2): 72–96. http://www.iijsr.org/data /frontImages/gallery/Vol._1_No._2/6.pdf.

CHAPTER 6

The Urgent Need for Sector Reforms: The Case of the Arab Republic of Egypt

The electricity sector in the Arab Republic of Egypt is led by the Ministry of Electricity and Renewable Energy (MoERE), established in the early 1960s, with the main mandate of securing electricity supply at the national level. A reform in the early 2000s resulted in the corporatization of the power sector into an Egyptian joint stock (holding) company: the Egyptian Electricity Holding Company (EEHC). Following this reform, a legal unbundling of generation, transmission, and distribution assets took place: six generation utilities (GUs), nine distribution utilities (DUs), and the Egyptian Electricity Transmission Company (EETC) were created, all of which are 100 percent owned by EEHC. Additionally, three private GUs were established under 20-year build-own-operate-transfer (BOOT) contracts with EETC, which since 1996 has operated thermal power plants with a combined installed capacity representing 6.4 percent of Egypt's total installed capacity of 32 gigawatts (GW).[1] Figure 6.1 shows the current structure of the Egyptian electricity sector.

Since 2001, the sector's regulation has been mandated to the Electric Utility and Consumer Protection Regulatory Agency (EgyptERA), which regulates, supervises, and controls electric-power-related activities, including generation, transmission, distribution, and consumption. EgyptERA's mission is to ensure electricity supply, quality, and access at equitable prices, while considering environmental issues.

In the current electricity market structure, EETC acts as single buyer and is the only utility licensed for extra high voltage (EHV) and high voltage (HV) electricity transmission. The EETC purchases electrical energy from the six GUs, the three private ones, and a small independent power producer (IPP) as well as from the New and Renewable Energy Authority (NREA). It then sells the electricity to the nine DUs and to approximately 100 EHV and HV consumers. In addition, the EETC conducts energy sales and exports with neighboring economies over the existing interconnections.

Figure 6.1 Electricity Sector Organization, Arab Republic of Egypt

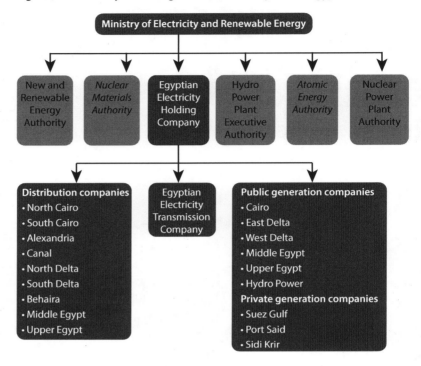

Electricity Generation

The GUs produce electricity, which is sold to EETC, and are responsible for the management, operation, construction, rehabilitation, and overhauling of power plants. As shown in table 6.1, out of 32,015 megawatts (MW) of installed capacity, thermal power plants represent 89 percent, while hydropower and renewable energy (wind and solar) represent 9 percent and 2 percent, respectively. Figure 6.2 indicates that the technology most often used is steam (43 percent), followed by combined cycle (35 percent) and gas (11 percent).

Table 6.1 Generation Mix, Arab Republic of Egypt, 2013

Generation type	Amount
Hydropower generation (MW)	2,800
Thermal power generation (MW)	26,480
New and renewable energy (wind and solar) (MW)	687
Private sector BOOTs (thermal) (MW)	2,048
Total installed capacity (MW)	**32,015**
Total generated energy (GWh)	**168,050**

Source: EEHC 2013/14.
Note: BOOT = build-own-operate-transfer; GWh = gigawatt-hours; MW = megawatts.

Figure 6.2 Share of Technology Type in Generating Electricity, Arab Republic of Egypt, 2013
Percent

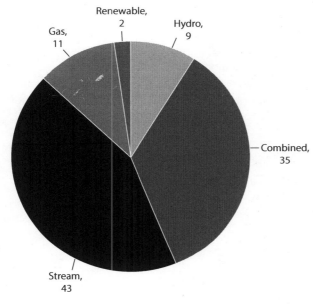

Source: EEHC 2013/14.

Electricity Transmission

EETC is the single public entity responsible for managing, operating, and maintaining the electric transmission grid on EHV and HV levels across Egypt. Table 6.2 includes basic data on the transmission lines and substations of the transmission utility (TU).

Egypt's geographical position allows for electricity exchanges to take place through existing regional interconnections, namely with Libya and Jordan. In 2013, Egypt exported 460 gigawatt-hours (GWh), almost eight times more than the amount it imported (61 GWh) (EEHC 2013/14). An electrical interconnection between Egypt and Saudi Arabia is currently under implementation, and the possibility of Egypt–Sudan and Egypt–Ethiopia–Sudan connections is under study.[2]

Table 6.2 Electricity Transmission Data, Arab Republic of Egypt, 2013

Transmission	Amount
Total transmission lines and cables (132 kV, 220 kV, 500 kV) km	44,213
High voltage (66 kV and 33 kV) substation capacity MVA	99,635

Source: EEHC 2013/14.
Note: km = kilometers; kV = kilovolts; MVA = megavolt ampere.

Electricity Distribution

The nine DUs responsible for the distribution and sale of electric energy purchased from EETC sold a total volume of 120,826 GWh to 30.6 million customers in 2013. Of this energy volume, 51.3 percent was distributed to the residential sector (see figure 6.3), which represents 73 percent of all medium- and low-voltage customers. The DUs are also responsible for managing, operating, and maintaining the medium- and low-voltage grid, as well as for preparing, for instance, forecasts of customer demand. The share of private DUs does not exceed 1 percent of the market. Table 6.3 includes basic data for the distribution lines and substations.

Figure 6.3 Energy Sold from Distribution Utilities by Sector (medium- and low-voltage consumers), Arab Republic of Egypt, 2013

Percent

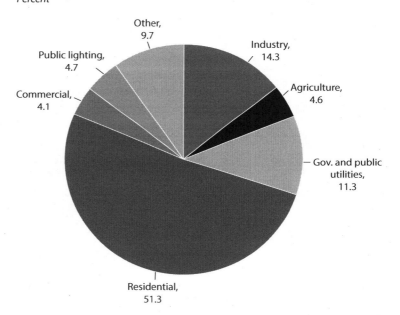

Source: EEHC 2013/14.

Table 6.3 Electricity Distribution Data, Arab Republic of Egypt, 2013

Transmission	Amounts
Distribution transmission lines length (km)	425,611
Distribution substation capacity (MVA)	64,956
Customers	
Number of consumers (millions)	30.6

Source: EEHC 2013/14.
Note: km = kilometers; MVA = megavolt ampere.

Comparison of Egyptian Generation Utilities

The unbundled nature of Egypt's electricity sector makes it possible to conduct a comparative performance assessment of both GUs and DUs. Table 6.4 compares five public Egyptian GUs against one another through a set of characteristic and performance indicators.[3] Comparisons are also made with Middle East and North Africa (MENA) median values when available.[4]

The first set of indicators listed in table 6.4 characterizes the type and size of GUs: all of which are big, at above 1 GW. Egyptian GUs have high installed capacities ranging from 3.4 GW for the Upper Egypt Electricity Production Company (UEEPC) to 6.2 GW for the Cairo Electricity Production Company (CEPC). The utilities mainly operate thermal power plants (including gas, steam, and combined cycles) and tend to be highly staffed (the West Delta Electricity Production Company [WDEPC] has the largest number of employees, at 8,577).

The capacity factor indicates how much of the plants' potential capacity was used during the year. This factor indicates that units were working between 57 percent and 70 percent of their full capacity: although Cairo, East Delta, and West Delta Electricity Production Companies have capacity factors similar to the MENA median (58 percent), GUs in Middle Delta and Upper Egypt have higher figures. The availability factor—that is the percentage of a total year that plants were in service—is similar, and ranges from 79 percent to 87 percent, with the exception of UEEPC, which has an availability factor of 91 percent. This indicator depends on generation outages, whether they are caused by failure, maintenance, or the availability of fuel.

All Egyptian GUs have values of operating expenses (OPEX) per employee lower than the MENA median ($297,000); Cairo and Upper Egypt GUs' values are about half this median, whereas the other three GUs have significantly lower values (and a higher number of employees). Given that all these GUs operate with similar types of technologies and fuel use, a lower ratio could imply overstaffing and therefore greater labor inefficiencies.

The cost structure indicators show that most OPEX is for fuel and lubricant (ranging from 79 percent to 88 percent), rather than labor. The share of fuel in OPEX is lowest for the Middle Delta Electricity Production Company (MDEPC) (79 percent) and could be a direct consequence of the generation technology used by this utility, which is essentially combined cycle. This utility's relatively low fuel expenses could also explain why it has one of the highest capacity factors (65 percent). CEPC has among the highest OPEX values, yet the second-lowest labor cost as a share of total OPEX (8 percent). These figures are consistent with the fact that the three Delta GUs have more staff than the others.

The cost-recovery indicators show that none of the GUs in Egypt recover their total OPEX or their total costs from sales, with the exception of MDEPC, which recovers its OPEX but not its total costs. The GU in Upper Egypt, on the other hand, has the lowest recovery rates both of OPEX (59 percent) and

Table 6.4 Comparing the Performance of Generation Utilities across Indicators, Arab Republic of Egypt, 2012/13[5]

	Indicator name	Unit	Cairo	East Delta	West Delta	Middle Delta	Upper Egypt	Median MENA
General	Installed capacity	GW	6.2	5.9	5.0	4.8	3.4	—
	Net generation	TWh	31	31	25	27	21	—
	Employment	employees, thousands	5.4	7.0	8.6	6.2	3.2	—
	Technology type	%	Gas (10), steam (48), CC (41)	Gas (42), steam (38), CC (20)	Steam (77), CC (18), steam (5)	Steam (8), CC (92)	Steam (56), CC (44)	—
	OPEX	$ millions	752	647	625	470	587	—
Technical and operational	Capacity factor	%	58	60	57	65	70	58
	Availability factor	%	82	87	84	79	91	93
	OPEX/employee	$ thousands	138	91	77	76	179	297
Financial (Cost structure)	Share of cost of fuel, lubricant in total OPEX	%	88	88	81	79	83	75
	Share of labor cost in total OPEX	%	8	10	15	12	5	12
Financial (Cost recovery)[a]	Energy sales/total OPEX	%	91	97	76	139	59	109
	Energy sales/total costs[b]	%	61	68	56	83	46	107
Financial (Balance sheet)	Accounts receivable	Days	412	222	603	274	571	40
	Debt/equity	%	—	3,484	3,074	2,509	1,270	357
	Current assets/current liabilities	%	52	37	67	68	56	95
Financial (Profitability)	Return on assets	%	0.02	0.02	0.01	0.03	0.35	3.00
	Return on equity	%	0.6	0.3	0.1	0.4	3.0	7.0

Source: MENA Electricity Database.

Note: CC = combined cycles; GW = gigawatts; MENA = Middle East and North Africa; OPEX = operating expenses; TWh = terawatt-hours; — = not available.
a. The values of MENA medians above 100 percent are mainly driven by Omani generation utilities (12 of 23 used in this study), which have 193 percent and 112 percent median values, respectively, for the two cost-recovery indicators.
b. Data from regulator.

total costs (46 percent). According to EgyptERA, in 2013, a total of $1.6 billion was provided to the five GUs in the form of government subsidies. When OPEX are not being recovered, this could indicate that electricity is being underproduced, resulting in insufficient sales. But Egyptian utilities have high availability and capacity factors. Another possibility could be the high cost of fuel for generation, yet Egyptian fuel is in fact highly subsidized. The inability of GUs to recover their costs must therefore be from low tariffs.

The accounts receivable of Egyptian GUs are very high, from six to 15 times higher than the MENA median. But the average number of days involved could hamper the recovery of OPEX from sales. In the management of day-to-day activities, these delays can cause cash shortfalls, causing, for example, deferrals of scheduled maintenance. In Egypt, compromising maintenance activities might not be an issue, because plant availability factors are already quite high. However, there seems to be a relation between accounts receivable delays and low OPEX recovery from sales.

The debt-to-equity ratios of Egyptian GUs are extremely high, from 1,270 percent to 3,484 percent. This is between four and 10 times higher than the MENA median of 357 percent. Unable to recover costs, utilities most likely make use of debt instruments to finance their activities, or at least to cover their operating costs. Not surprisingly, when it comes to mobilizing liquid assets to repay short-term debts, Egyptian GUs are also underperforming, at below 70 percent on average for all utilities.

Based on the financial stance of these utilities, it comes as no surprise that the return on assets (ROA) and return on equity (ROE) figures are close to 0 percent. The median value for ROA and ROE in the MENA region is 3 percent and 7 percent, respectively, whereas the most profitable utility, UEEPC, has ROA and ROE values of 0.35 percent and 3 percent, respectively.

Comparison of Egyptian Distribution Utilities

Table 6.5 compares nine Egyptian DUs across a set of performance indicators. The right-hand column presents median values for MENA DUs, thus allowing for a broader comparison beyond Egypt.

Egyptian DUs perform technically well when considering two indicators: the load factor and distribution losses. Both have values close to the regional MENA median except for the Canal Electricity Distribution Company (CEDC), with a low load factor of 38 percent. High load factors in general lead to lower distribution losses, yet this is not necessarily the case observed in Egypt. This could be a result of high nontechnical losses (due to theft and erroneous meter readings), which contribute about 25 percent on average to total distribution losses.[6] If the losses could be improved by 1 percentage point, this would result in a savings of 1,626 GWh per year, equivalent to about $71 million.[7] A potential area of improvement would therefore be the reduction of nontechnical distribution losses, which could be a way of increasing the ROE of the utilities. Currently these ROE values are very low, as shown in table 6.5.

Table 6.5 Comparing the Performance of Distributors across Indicators, Arab Republic of Egypt, 2012/13

Category	Indicator name	Unit	North Cairo	South Cairo	Alexandria	Canal	North Delta	South Delta	El Behera	Middle Egypt	Upper Egypt	MENA Median
Technical and operational	Load factor	%	62	64	61	38	69	60	62	69	68	60
	Distribution losses	%	10	8	11	6	9	10	10	11	8	10
	OPEX/employee	$ thousands/employee	46	47	24	47	39	28	35	37	38	188
	OPEX/connection	$/connection	160	169	134	230	101	75	157	115	119	346
	OPEX/kWh sold	$/kWh	0.04	0.04	0.04	0.04	0.03	0.03	0.04	0.03	0.03	0.1
	OPEX/km	$ thousands/km	12	14	15	10	9	10	9	6	6	19.6
Commercial (Consumption and billing)	Total billing/connection	$/connection	138	148	111	197	97	68	132	96	101	299
Financial (Cost structure)	Share of labor cost in total OPEX	%	21	21	41	20	24	35	26	27	26	12
Financial (Cost-recovery)	Energy sales/OPEX	%	88	87	83	86	96	91	84	84	85	93
	Energy sales/total costs	%	84	82	—	80	86	83	75	75	73	88
Financial (Balance sheet)	Accounts receivable	days	188	293	82	62	256	79	186	117	182	121
	Debt/equity	%	850	1,282	—	685	677	523	527	501	571	523
	Collection rate	%	93	86	99	94	84	93	95	92	88	93
	Current assets/current liabilities	%	71	81	77	66	97	103	103	85	113	85
Financial (Profitability)	Return on assets	%	0.19	2.6	0.18	1.87	0.3	0.23	0.04	0.06	0.06	3.00
	Return on equity	%	0.6	8.8	0.3	7.7	0.8	0.5	0.1	0.1	0.2	7

Source: World Bank calculations.

Note: km = kilometer; kWh = kilowatt-hours; MENA = Middle East and North Africa; OPEX = operating expenses; — = not available.

The OPEX per employee across Egypt's DUs is much lower than the MENA median of $188,000. This is because they have the largest number of employees in the region, ranging from 8,083 for the Upper Egypt Electricity Distribution Company (UEEDC) to 17,917 for the South Cairo Electricity Distribution Company (SCEDC). By comparison, the DU with the largest number of employees outside Egypt is Morocco's Lyonnaise des Eaux de Casablanca (LYDEC), with 3,850, which covers electricity, water, and sanitation for the region of Casablanca. Meanwhile, OPEX per employee varies across DUs, ranging from $24,000 for the Alexandria Electricity Distribution Company (AEDC) to $47,000 for SCEDC.

The high cost of overstaffing is evident in labor's share of total OPEX—ranging from 20 percent for CEDC to 41 percent for AEDC. This is two to three times higher than the MENA median. Moreover, this is despite the relatively low cost of labor in Egypt.

On the commercial front, OPEX per connection oscillates between $75 for SDEDC to $230 for CEDC. These are low values when compared to the MENA median. For total billing per connection, South Delta has the lowest value and Canal Electricity the highest. In the case of CEDC, a high OPEX per connection could indicate that interruptions in supply are regularly and promptly solved. This would lead to customers having a continuous supply of electricity and, hence, allow their consumption to be high. Another reason for this high OPEX per connection and high billing rate might be that the utility spends money and effort in tracking bills and ensures that collection is frequent. It is also interesting to observe that the OPEX to sell a unit of energy in Egyptian DUs is two to three times higher than the MENA median.

All Egyptian DUs boast high collection rates; six out of nine utilities are close to or above the regional median value of 93 percent. Yet in almost all cases, collection periods are long (at 62 days, CEDC's is the shortest among Egypt's DUs; North Cairo's is the longest, at almost 6 months). This can be attributed to delayed collection cycles resulting from the time-consuming manual registration of readings and bills. The EEHC has been exploring the option of shifting to smart meters since 2013 as a way to reduce both nontechnical losses and the time involved in bill collection.

High debt-to-equity ratios and low current ratios suggest that the Egyptian DUs are not financially independent and rely heavily upon debt-financing instruments and mechanisms. Although this is a common trend among DUs in the MENA region (the median value of debt-to-equity is 523 percent and for the current ratio, 85 percent), Egyptian utilities face financial constraints largely due to tariff levels, which did not allow full costs to be recovered from sales in 2013. Revenues of DUs in Egypt also included elements beyond pure electricity sales, as was the case for CEDC, which also included subsidies for the electricity exported to Gaza. None of the Egyptian DUs recover their OPEX from sales, and at best recover only 86 percent of their total costs through sales. Profits, on the other hand, are positive when considering revenues from electricity sales as well as other sources, but remain negative if only electricity sales are considered, explaining why the DUs in table 6.5 all have positive ROA and ROE.

Evolution of Egypt's Electricity Sector since 2014

The electricity sector in Egypt has gone through a number of changes since 2014 that are worth mentioning, given that this analysis is based on 2013 data.

In 2014, Egypt embarked on an ambitious energy subsidy reform and laid out its plans to phase out subsidies within five years to reach 0.5 percent of gross domestic product (GDP) by 2019, with the remaining subsidies covering only liquefied petroleum gas (LPG) and electricity consumption of the poorest households. The fiscal burden of Egypt's energy subsidies had grown continuously over the two decades up to 2014: the budget share of energy subsidies increased from 9 percent to 22 percent between 1990 and 2014. Electricity prices have risen cumulatively over the past three years, by more than 85 percent across consumer categories, and fuel prices have been raised twice, ranging from an accumulated increase of 60 percent to 150 percent across different fuel products from 2015 to 2017. Three successive electricity tariff increases and two major petroleum price reforms since 2014 have reduced energy subsidies from almost 7 percent to around 2.6 percent of GDP between 2014 and 2017 (as projected).

In the electricity sector, the process is led by EgyptERA, the electricity regulator. For the years 2018 and 2019, it is planned that the regulator will present Egypt's Cabinet with (a) the current average electricity tariff charged to consumers; (b) an estimate of the average electricity tariff consistent with cost-recovery based on actual fuel costs, fuel mix, and foreign exchange costs applicable in each year; and (c) an estimate of the average electricity tariff consistent with electricity subsidy targets. Based on these inputs by EgyptERA, the Cabinet would decide the average tariff, and the board of EgyptERA would approve the associated tariff structure to be issued by a ministerial decree. This institutional process, which strengthens the position of the regulator beyond what it was, is underpinned by the new Electricity Law No. 87/2015 and supporting executive regulations and has been successfully piloted during the tariff revision for 2017, enabling the regulator to raise tariffs beyond the original five-year trajectory.

The energy sector is being prioritized for governance reforms due to its higher institutional capacity. In the electricity sector, the MoERE has decided to set up a modern governance structure for new generation assets, with a separate company for each of the three 4.4 GW combined cycle gas plants under construction, and using international norms for staffing and skills. Other initiatives include (a) setting up an internal audit department in the EEHC for the first time; (b) publishing the methodology for determining electricity tariffs across consumer categories for the first time, based on Cabinet approval; (c) initiating a business planning framework for all sector entities; and (d) implementing the decision of the EgyptERA to conduct public hearings on key policy issues from 2018.

Following the electricity shortages of summer 2014, the sector has advanced a major investment program aimed at improving the security of supply. However, significant inefficiencies remain in both the dispatch of the generation plant and the operation of the transmission and distribution networks. The new Electricity Law No. 87/2015 (DPF 1 Prior Action 1.5) envisages a full modernization of the sector. Its provisions strengthen the authority and transparency of the regulator and provide for an eight-year transition toward a competitive market. A critical first step is the separation of the EETC from its current role as a subsidiary of the EEHC, to become a network operator independent of generation and distribution activities, improve transparency and accountability of state-owned entities, promote competition and private investment in the sector, and provide nondiscriminatory third-party access to the grid.

Egypt has barely begun to develop its rich renewable energy resources, which include excellent conditions for commercially viable wind power as well as high-intensity direct solar radiation throughout its territory. Egypt's early investments in renewable energy were government owned; however, its ambitious plans to double the share of its generation capacity coming from renewable sources to 20 percent by 2022—and thus reduce reliance on fossil fuels—call for a substantial scale-up in private investment. The new renewable energy law (no. 203/2014) reduces risks and improves the financial viability of investments in wind power and solar photovoltaics (PV), improving the climate for private sector investment. The law and its associated feed-in tariff regulations provide incentives for the first 4,300 MW (wind and solar PV) as well as a regulatory framework for further private investment through competitive bidding mechanisms for IPPs. Moreover, recent increases in grid electricity tariffs combined with declining costs of renewable energy are increasingly making renewable energy solutions more competitive from an end-user perspective.

Conclusion

In this chapter, we analyzed the performance of Egyptian electricity utilities in 2013. Egypt has some serious challenges to overcome in the electricity sector, and maximizing the efficiency of its utilities can help achieve this. The period 2011–14 witnessed two peaks of political instability, which makes the issue of satisfying the increasing electricity demand a highly sensitive one, in particular when three-quarters of all the electricity volume sold by the DUs is destined for the residential sector. Although thermal power plants represent almost 90 percent of total capacity in the country, the sharp decline in oil and natural gas production—changing Egypt's status from a net exporter to a net importer—makes electricity supply even more challenging.

The five GUs studied in this chapter are big (3 GW to 6 GW of installed capacity) and rely on thermal power plants (gas, steam, and combined cycles). Although they perform reasonably well from a technical standpoint, Egyptian

GUs do not recover their OPEX (with the exception of the Middle Delta Electricity Distribution Company), although Cairo and East Delta are not far from doing so. GUs have financial performance indicators that are of real concern when compared to the rest of the region. In addition, overstaffing appears to be another area of concern, particularly in the three Delta GUs.

The nine Egyptian DUs are big as defined by this study (that is, more than 2 million connections) except for the El-Behera Electricity Distribution Company (EEDC), which is of medium size. Again, overstaffing appears to be a key area of concern: for example, labor's share of total OPEX is two to three times bigger than the MENA median. The OPEX per unit of energy sold is three to four times bigger than the median of the rest of the region. No Egyptian DU recovers its OPEX from sales, but all values are above 80 percent. Most balance sheet indicators show poor performance (though to a degree not nearly as concerning as that of GUs). Finally, the profitability of DUs is low: ROA and ROE tend to be low, with the notable exception of ROE for both South Cairo Electricity Distribution Company and CEDC, which are above the MENA median.

EEHC's expansion plan indicates that 3,000 MW of additional electricity generation capacity would need to be added every year to meet 2020's forecasted demand. What this chapter has shown is that simply expanding supply will not be sufficient to improve the performance of the Egyptian electricity sector. The financial situation of Egyptian GUs is so delicate that financial restructuring will presumably be needed (for example, by utilities raising the companies' equity through conversion of the public debt into equity). In addition, tariff reforms are required if the sector is to be financially viable. Improved efficiency of electricity operators and better corporate governance will inevitably be part of the solution to the sector's challenges.

Last but not least comes the issue of data collection and its quality. Most of the quantitative evidence on performance provided in this chapter is not available online, and required numerous exchanges with the regulator to check the validity of figures and establish a common understanding of the factors behind specific values. Even after these efforts, some values were left aside as they did not appear to be reasonable. The exercise of carrying out periodic performance assessments with the direct involvement of the GUs and DUs should be reinforced by the regulator. The multiplicity of genera- tion and distribution actors within the same economy provide a unique opportunity to benchmark performance across operators. But for these exer- cises to be of use, operators need to adopt international accounting standards and a cost accounting system (it remains unclear which Egyptian utilities have done so) and DUs should implement supervisory control and data acquisition (SCADA) (absent across Egypt's DUs). Without these, the reli- ability of many of the financial and commercial indicators collected is deemed to be very low.

Notes

1. These three private generation utilities have not been included in this study.
2. Within the framework of the Eastern Africa Power Pool and Nile Basin Initiative Plans.
3. Other Egyptian generation utilities—namely the publicly owned Hydro Power Plants Electricity Production Company (2,800 MW) and private thermal generation—are not included in this analysis because data were not collected.
4. We had insufficient data for non-MENA generation utilities to make a meaningful comparison.
5. The indicators energy sales/total OPEX, energy sales/total costs, and accounts receivable are not applicable to generation utilities based on how these indicators were categorized for the purpose of the MENA Electricity Database. However, for comparative purposes, their values are presented and discussed in this chapter but not in previous chapters of this book.
6. Based on a calculation of seven distribution utilities in Egypt.
7. Savings of Egyptian LE 486 million (EEHC 2014).

References

EEHC (Egyptian Electricity Holding Company). 2014. *Annual Report 2013/2014*. http://www.moee.gov.eg/english_new/EEHC_Rep/REP-EN2013-2014.pdf.

EgyptERA (Electric Utility and Consumer Protection Regulatory Agency). Cairo: EEHC. http://www.egyptera.org.

CHAPTER 7

Harvesting Results from a Restructuring of the Power Sector: The Case of Jordan

In 1994, the government of Jordan initiated an electricity restructuring and reform program that opened the sector to private sector involvement. Today the electricity sector is unbundled, and sector policy is set by the Ministry of Energy and Mineral Resources (MEMR). Generation utilities (GUs), whether public or private, sell their electricity to a single buyer, the National Electric Power Company (NEPCO), which is also the fuel purchaser bearing all costs and risks related to fuel price fluctuations. NEPCO also acts as a transmission system operator (TSO), manages the nation's electricity transmission infrastructure, and sells electricity to the three main distribution utilities (DUs): the Jordan Electric Power Company (JEPCO), Irbid District Electricity Company (IDECO), and Electricity Distribution Company (EDCO). In addition to the Central Electricity Generation Company (CEGCO) and Samra Electric Power Generating Company (SEPCO), four independent power producers (IPPs) are also present: AES Jordan (the first IPP in Jordan), the Qatrana Electric Power Company (QEPCO), the Amman Asia Electric Power Company (AAEPC), and AES Levant. Figure 7.1 shows the organization of the Jordanian electricity sector as of 2014.

The introduction of the General Electricity Law No. 64 in 2002 marked an important milestone. Soon after, the Electricity Regulatory Commission (ERC) was established as an autonomous regulatory body tasked with licensing the country's electric utilities (generation, transmission, and distribution). In 2014, its mandate was expanded to include regulation of other forms of energy—namely nuclear and mining activities—and it is now known as the Energy and Minerals Regulatory Commission (EMRC). Another milestone was the introduction of the Renewable Energy Law in 2012. According to this law, self-production is authorized by law, with the possibility of either net metering or selling excess electricity to the grid.

Figure 7.1 Electricity Sector Organization, Jordan, 2014

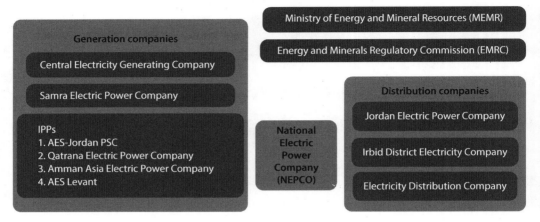

Source: World Bank.
Note: IPP = independent power producer.

Table 7.1 Generation Mix, Jordan, 2013/14

Generation type	Amount (MW)
Hydropower generation	12.0
Steam	791.0
Diesel	27.0
Gas	621.0
Combined cycle	1,737.0
Wind	1.4
Biogas	3.5
Total installed capacity (MW)	3,193.0
Total generated energy (GWh)	17,886.0

Source: NEPCO 2013.
Note: GWh = gigawatt-hours; MW = megawatts.

Electricity Generation

In 2013, thermal power plants represented 99 percent of the installed capacity of GUs. Hydropower and renewable energy made up the remaining 1 percent. Table 7.1 presents Jordan's installed capacity by technology. Until 2010, around 80 percent of the electricity generated was from natural gas imported through the Arab gas pipeline. Frequent interruptions in supply led to the use of costlier secondary fuels for generation, that is, diesel oil and heavy fuel oil (HFO).

The GUs, however, are shielded from the risk of fuel costs by NEPCO, to which they also sell all the electricity produced. Figure 7.2 illustrates the percentage share of each fuel from 2009 to 2013, clearly showing that the energy mix of Jordan shifted in these years, reaching more than 90 percent natural gas and more than 75 percent of HFO and diesel by 2013.

Figure 7.2 Share of Fuel Type in Electricity Generation, Jordan, 2009–13

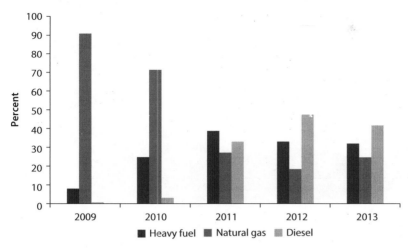

Source: NEPCO 2012, 2013.

Electricity Transmission

NEPCO is the state-owned single buyer of all electricity produced in Jordan except for renewable energy sources that are directly connected to the distribution network. NEPCO also has the role of system operator and is responsible for managing and operating the Jordanian electricity transmission grid, which consists of 132 kilovolts (kV) and 400 kV networks. Table 7.2 shows data related to the transmission network length and substation capacities.

The transmission system it operates interconnects the power generation plants with the load centers. The total length of the transmission network is about 4,463 kilometers (km) of circuit and includes 400 kV tie lines with Syria.

Table 7.2 Electricity Transmission Data, Jordan, 2013

Transmission	2013
Total transmission lines and cables (132 kV and above) (km)	4,463
High voltage (132 kV and 33 kV) substation capacity (MVA)	7,444

Source: NEPCO 2013.
Note: km = kilometers; kV = kilovolts; MVA = megavolt ampere.

Electricity Distribution

Three DUs operate in Jordan, each covering a certain geographical region. Although JEPCO's service area includes industrial areas as well as the capital city of Amman, the other two DUs cover mostly rural areas. The total amount of electricity sold in 2013 amounted to 13.8 terawatt-hours (TWh), of which 62 percent was sold by JEPCO, which has a customer base of over a million customers. Table 7.3 lists basic data for the distribution lines and substations.

As in most of the economies in the Middle East and North Africa (MENA) region, most of Jordan's electricity is used in the residential sector, as shown in figure 7.3.

Table 7.3 Electricity Distribution Data, Jordan, 2013

Distribution	Amount
Distribution transmission lines, km	57,635[a]
Distribution substation capacity, 400/132/33 MVA	3,760
Customers	
Number of consumers	1,744,000

Source: NEPCO 2013.
Note: EDCO = Electricity Distribution Company; IDECO = Irbid District Electricity Company; JEPCO = Jordan Electric Power Company; km = kilometers; MVA: megavolt ampere.
a. Sum of length of distribution network of EDCO, JEPCO, and IDECO, according to the MENA Electricity Database.

Figure 7.3 Volume of Energy Distributed by Sector, Jordan, 2013
Percent

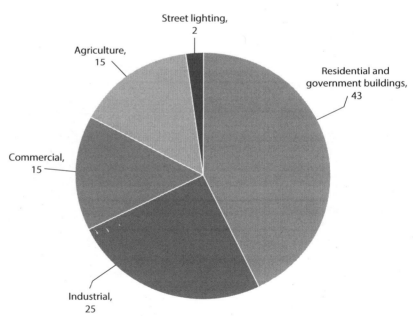

Source: NEPCO 2013.

Electricity Tariffs between Utilities

The main responsibility of EMRC is to set tariffs ensuring that the prices charged by licensees are sufficient to finance their activities and allow them to earn sufficient return on their investments. As far as DUs are concerned, EMRC sets both the tariff between them and end consumers and the tariff that they pay to NEPCO. As far as private GUs are concerned, the tariff that NEPCO pays them is set once the generator enters the market in accordance with a power purchase agreement (PPA). CEGCO's tariff was determined at the time of its privatization. The other GUs—that is, the IPPs, in the current single-buyer model—compete for the market after NEPCO specifies the capacity and energy needs, location, and time when capacity is required. The winner is the generator with the lowest levelized price. For generator plants, the cost of fuel is passed through, and NEPCO provides the fuel (or pays its cost). The mechanism is different for the only state-owned GU, SEPCO, for which the tariff is decided by EMRC following a request by SEPCO and NEPCO.

Comparison of Jordanian Generation Utilities

Six GUs in Jordan are compared against one another in table 7.4, as well as to the MENA region median.[1] Although most of the data defining the general characteristics of GUs is available, this is not always the case for the technical and operational, commercial, and financial indicators. Most of the data gaps concern the IPPs, from which it was a challenge to obtain data.

Jordanian GUs are thermal power plants running on HFO, natural gas, and diesel. The sizes of the six Jordanian GUs are heterogeneous: AES Levant, AES PSC, and QEPCO are considered small in this study (less than 500 megawatts, MW); AAEPC, medium (500 MW to 1 gigawatt, GW); and CEGCO and SEPCO, big (above 1 GW).

Looking at the ratio of installed capacity to employees or the ratio of operating expenses (OPEX) to employees, CEGCO, AAEPC, and SEPCO appear to be overstaffed. The result for CEGCO could be explained by the fact that it is the oldest and largest GU in the country, and high staff numbers could be customary. Another reason could be the ownership structure of CEGCO: the private sector owns 51 percent of the utility's shares, the public sector owns 49 percent (of which the Government of Jordan owns 40 percent, and the social security corporation owns 9 percent). This argument is strengthened by the fact that the IPPs have a much lower staff number than publicly owned utilities: 47 for AES Levant, 51 for AES PSC, 75 for QEPCO, and 287 for AAEPC. In the case of AAEPC, the plant was not fully operational in 2013, and most of the employees numbered here might have been those outsourced during the precommissioning phase. According to the MENA Electricity Database (MED), AAEPC had 135 outsourced employees and 152 full-time employees in 2013.

The utilities with the lowest capacity factors are the two IPPs that were not fully operational in 2013—AES Levant and AAEPC. The high capacity factors for AES PSC and QEPCO, on the other hand, could be explained by the fact that they are both private utilities and have to fulfill contractual obligations regarding sales to NEPCO under their PPAs. In the case of CEGCO, several old generating units have retired, and, due to regulatory constraints, the utility does not, in principle, have the right to add new units, which could explain the low capacity factor.[2]

Table 7.4 Comparing the Performance of Generation Utilities across Indicators, Jordan and MENA Median, 2013[3]

Category	Indicator name	Unit	AES Levant[a]	AAEPC[a]	AES PSC	CEGCO	QEPCO	SEPCO	MENA median
General	Installed capacity	MW	246	573	370	1,687	373	1,031	—
	Net generation	TWh	0.6	0.3[b]	2.6	7.4	2.4	4.5	—
	Employment	Employees	47	287	51	1,037	75	345	—
	Fuel mix	n.a.	HFO, natural gas	HFO, natural gas, diesel	Natural gas	Natural gas, diesel, HFO, small hydro, wind	Natural gas, diesel	Natural gas, diesel	—
Technical and Operational	OPEX	$, millions	—	89	472	1,401	440	833	—
	Capacity factor	%	28	6.3	80	50	75	50	58
	Availability factor	%	99	—	—	90	98	—	93
	OPEX/employee	$, millions/employee	—	0.3	9.3	1.3	5.9	2.4	0.3
Financial (Cost structure)	Share of cost of fuel, lubricant in total OPEX	%	—	63	98	94	99	97	75
	Share of labor cost in total OPEX	%	—	1.3	0.7	1.3	—	0.7	12.0
Financial (Cost recovery)[c]	Energy sales/total OPEX	%	—	20	68	99	75	101	109
	Energy sales/total costs	%	—	14	—	96	71	95	107
Financial (Balance sheet)	Accounts receivable	Days	—	—	62	98	91	50	40
	Debt/equity	%	—	290	333	354	621	876	357
	Current assets/current liabilities	%	—	123	287	95	488	113	95
Financial (Profitability)	Return on assets	%	—	—	—	12	5	4	3
	Return on equity	%	—	—	36	21	25	17	7

Source: World Bank calculations.

Note: AES Levant = AES Levant Holding BV Jordan PSC; AAEPC = Amman Asia Electric Power Company; AES PSC = Amman East Power Plant; CEGCO = Central Electricity Generation Company; HFO = heavy fuel oil; MENA = Middle East and North Africa; MW = megawatts; n.a. = not applicable; OPEX = operating expenses; QEPC = Qatrana Electric Power Company; SEPCO = Samra Electric Power Generating Company; TWh = terawatt-hours; — = not available.

a. Fully operational only in 2014.

b. Not fully in operation in 2013, which explains the low electricity output and capacity factor.

c. The values of MENA medians above 100 percent are mainly driven by Omani generation utilities (12 out of 23 used in this study), which have high values, as shown by the median values of 193 percent and 112 percent for the two cost-recovery indicators.

Because Jordan is one of the only economies in the region not to have subsidies for fuel, OPEX values[4] are much higher than those observed elsewhere in the region. In fact, the majority of GUs' spending is on fuel, ranging from 95 percent to 99 percent. Consequently, the share of labor costs in OPEX is considerably smaller than the MENA median (12 percent). CEGCO has the largest number of employees, yet its OPEX per employee, estimated at $1.3 million, is seven times the MENA median value of $267,000 per employee. AES PSC has the highest OPEX per employee value in Jordan, at $9 million. In the case of AAEPC, which was not yet fully operational in 2013, this indicator might be misleading, particularly when considering the net generation figures in table 7.4, which show that AAEPC generated about 15 times less electricity than SEPCO.

For an installed capacity of 1,687 MW and net generation of 7.4 TWh, CEGCO compares poorly with SEPCO, which generated 4.5 TWh (equivalent to 60 percent of CEGCO's energy output), with three times fewer employees and almost two times less OPEX. The high fuel costs as a share of OPEX could be explained by analyzing the fuel mix of the two utilities.

SEPCO produced three times more electricity than CEGCO from natural gas sources (representing 25 percent of SEPCO's total generation). Natural gas is more efficient and could be the reason why SEPCO has lower OPEX even though the share of fuel costs in its OPEX is 70 percent. In the case of CEGCO, 72 percent of the electricity generated is from HFO, and because fuel costs constitute 94 percent of its OPEX, this explains why CEGCO also has the highest OPEX value among the GUs.

Only one of the utilities in table 7.4 recovered their total OPEX from energy sales: SEPCO. AAEPC is a private GU and is therefore expected to recover its OPEX, yet it has the lowest OPEX recovery rate. This could be explained by the fact that AAEPC was not fully operational in 2013, as can be observed in the listed electricity output of 317 gigawatt-hours (GWh), which is low for an installed capacity of 573 MW (compared with the 2,591 GWh output of AES PSC's 370 MW installed capacity). As far as full cost-recovery is concerned, although CEGCO and SEPCO are very close to fully recovering costs via energy sales, AES PSC and QEPCO are not quite there, and AAEPC is very far from cost-recovery (again, this could be because it was not fully operational in 2013).

We now look at financial performance. The debt-to-equity ratio in Jordan is similar across GUs and in some cases lower than the MENA median, with the exception of the fully state-owned utility SEPCO (876 percent). SEPCO has been expanding its generation capacity since 2010 by adding new generating units, which must have been financed mainly through debt rather than equity.

The GUs are all profitable, with high return on equity (ROE) and return on asset (ROA) values. In particular, the IPPs' ROE is twice the MENA median value of 7 percent. This is mainly a result of the provisions of the IPP contracts under which they operate and the PPAs with NEPCO that ensure that IPPs can actually make a profit. In addition, NEPCO absorbs the risk of fuel price fluctuations and so protects GUs from it.

Comparison of Jordanian Distribution Utilities

Table 7.5 compares the three DUs operating in Jordan against one another, using several performance indicators. The MENA median values are also included where available. While EDCO is considered a small DU in this study (less than 250,000 connections), IDECO and JEPCO are considered medium (250,000 to 2 million connections). Among the three DUs, EDCO (34 percent) has the lowest load factor, whereas IDECO has the highest (56 percent). The load factor depends upon the amount of electricity distributed and the peak load, which both vary according to the consumption patterns and type of consumers the DU serves.

Usually, the higher the load factor, the lower the distribution losses in the distribution system, and this is indeed the case for IDECO, whose distribution losses, at 11 percent, are the lowest of the three utilities. This value is similar to the MENA median value of 10 percent. The distribution losses of JEPCO are the highest, at 14 percent, because the utility distributes the largest amount of electricity and across a much longer set of medium- and low-voltage networks than the other two (27,000 km for NEPCO against almost 19,000 km and 12,000 km for IDECO and EDCO, respectively).[5]

OPEX per employee figures for Jordanian DUs are much higher than the MENA median, most likely due to the high OPEX figures attributed to other costs (labor costs represent only about 6 or 7 percent of OPEX for each utility listed in table 7.5).

EDCO covers the largest service area (68,359 square kilometers, km^2) and also has the smallest number of connections, which could explain the high OPEX costs per connection ($1,483), as well as the high OPEX per kilowatt-hour sold ($0.23). For IDECO, the values of OPEX per connection and OPEX per kilowatt-hour are the closest to the MENA median value, and closer than those of EDCO and JEPCO. It is the most efficient utility in terms of operational performance, also having the lowest OPEX per kilometer among the three DUs. However, JEPCO seems to underperform in this category (although this might not necessarily be the case), mainly as a result of its relatively high OPEX, which could be linked to the larger amount of energy purchased to supply to its consumers.

Total billing per connection is highest among customers serviced by EDCO ($1,528) and lowest for IDECO ($576). This could be the result of a difference in the tariffs applied by the two utilities. In the case of JEPCO, total billing per connection is high. This reflects high consumption in the capital, Amman, and also among the industrial consumers serviced by JEPCO, as compared with the more rural consumers serviced by the other two utilities (the agricultural sector represented only 5 percent of JEPCO's sales, whereas it represented 12 percent and 11 percent of sales for IDECO and EDCO, respectively).

Table 7.5 Comparing the Performance of Distributors across Indicators, Jordan and MENA Median, 2013

Category	Indicator Name	Unit	EDCO	IDECO	JEPCO	MENA median
Technical and Operational	Load factor	%	34	56	51	60
	Distribution losses	%	12	11	14	10
	OPEX/employee	$, thousands/ employee	230	197	448	188
	OPEX/connection	$/ connection	1,483[a]	547	1,038[a]	346
	OPEX/kWh sold	$/kWh	0.23[a]	0.10	0.14	0.1
	OPEX/km	$, thousands /km	26	12	43	19.6
Commercial (Consumption and billing)	Total billing/connection	%	1,528[a]	576	936[a]	299
Financial (Cost structure)	Share of labor cost in total OPEX	%	6	7	7	12
Financial (Cost recovery)	Energy sales/OPEX	%	97	107	93	93
	Energy sales/total costs	%	—	99	—	88
Financial (Balance sheet)	Accounts receivable	Days	117	120	122	121
	Debt/equity	%	1,476	981	576	523
	Collection rate	%	—	—	97	93
	Current assets/current liabilities	%	99	84	80	85
Financial (Profitability)	Return on assets	%	5	6	24	3
	Return on equity	%	16	20	12	7

Source: MENA Electricity Database.
Note: EDCO = Electricity Distribution Company; IDECO = Irbid District Electricity Company; JEPCO = Jordan Electric Power Company; km = kilometers; kWh = kilowatt-hours; MENA = Middle East and North Africa; OPEX = operating expenses; — = not available.
a. Outlier not used in calculations of MENA average and median values mentioned in earlier chapters of this book.

The only utility that fully recovers OPEX from sales is IDECO, although the values for EDCO and JEPCO are not far from cost-recovery. All Jordanian DUs have high debt-to-equity ratios, suggesting a high level of financing through debt. (This is common in the MENA region, where the median value is 523 percent.) The current shares for the three utilities are below 100 percent, with the highest value belonging to EDCO (99 percent). IDECO's low value could be attributed to the utility's low contribution of cash to current assets. JEPCO has the lowest current ratio, suggesting that the current liabilities are extremely high. Although JEPCO does have high receivables, this is unlikely to be the cause of the low current ratio because JEPCO collects its receivables within 120 days (comparable to the MENA median value) and has a high collection rate of 97 percent.

EDCO, JEPCO, and IDECO are profitable, as shown by the ROA and ROE indicators in table 7.5. Since the privatization of EDCO and IDECO in 2008, the revenues and profits of these utilities are controlled and regulated by the national regulator (EMRC), which reevaluates assets at the end of the concession period and at the time of obtaining a license. These licenses grant the companies a 10 percent profit on their regulatory asset base after the regulator reviews and approves their annual budgets, their projects, and anticipated electricity losses.

Evolution of Jordan's Electricity Sector since 2014

The electricity sector in Jordan has gone through a number of changes since 2014 that are worth mentioning, given that the analysis here is based on 2013 data.

The government of Jordan faces the challenge of pursuing its reform agenda while also accommodating an influx of Syrian refugees. An estimated 1.3 million Syrian refugees are currently residing in Jordan—equivalent to over 20 percent of Jordan's population before the start of the Syrian crisis in 2011. To mitigate supply risks while keeping up with demand, the Jordan 2025 strategy, approved in 2015, set targets to (a) increase the share of local energy sources in the energy mix (from 2 percent in 2014 to 39 percent by 2025); (b) reduce the energy intensity of the economy; and (c) decrease the percentage of electricity transmission and distribution losses (from 17 percent in 2014 to 11 percent by 2025).

The government's reform program aims to lock in the achievements of energy sector reforms over recent years despite the additional strain of the Syrian crisis and further strengthen resilience to external shocks of fuel supply interruptions and price volatility. Key measures under the government's multiyear reform program include (a) restoring the financial sustainability of the electricity sector, (b) diversifying gas import sources, (c) developing domestic energy resources, and (d) promoting energy efficiency.

The government succeeded in restoring the financial sustainability of the electricity sector by the end of 2015. The rising cost of fuels since 2010 had created a gross imbalance between costs and revenues for NEPCO. In 2013, the government adopted a five-year (2013–17) Electricity Tariff Adjustment Plan to restore the adequacy of NEPCO's revenue base. A number of factors allowed NEPCO to reach cost-recovery in the final quarter of 2015: tariff increases, a decline in international oil prices after mid-2014, a switch from oil to cheaper natural gas dating from mid-2015, and the commissioning of the first large-scale renewable energy plant. The government is committed to locking in its reform achievements through further tariff reforms with the aim of sustaining cost-recovery for NEPCO amid volatile energy import prices.

Jordan's private distribution sector comprises a number of bilateral performance agreements between the regulator and the DUs, with the aim of applying international best practices to achieve efficiency gains. Loss reduction targets for 2016 and 2017 were finalized by EMRC and agreed upon by the three DUs at the end of 2015, with a plan to agree on targets for 2018 and 2019 at the end of 2017.

NEPCO has developed a holistic strategy for securing a supply of relatively clean fuel. Implementation of the strategy began in 2015. The main thrust of the strategy is the diversification of supply sources. Natural gas, most of which is imported in liquid form through the liquefied natural gas (LNG) terminal in Aqaba, is sourced through two multiyear LNG supply contracts and on the spot. These contracts allowed NEPCO to provide natural gas for 84 percent of power generation until mid-2016. However, all of Jordan's long-term LNG imports remain linked to the Brent oil price, which makes the country vulnerable to price shocks. In addition to LNG, Jordan is pursuing longer-term supply options—including piped gas from the Arab Republic of Egypt, Iraq, and the Eastern Mediterranean—to ensure a secure and clean fuel supply to its electricity sector in the long term. Renewable energy is procured from IPPs. A total of 30 IPP projects, totaling 1,374 MW, are now at various stages of development. PPAs for around 1,000 MW of capacity have been signed and around 240 MW are operational. This makes Jordan a leader in private-sector-owned renewable energy in the MENA region.

Conclusion

This chapter analyzed the performance of Jordanian electricity utilities in 2013. Jordan's main challenge does not reside in energy access and infrastructure development but is primarily in guaranteeing supply to meet increased demand and as a net energy-importing country. The challenge is even greater since the Syrian refugee influx. Because its main fuel imports for electricity generation were interrupted in 2010, the country has had to look for alternative sources while depending on secondary options, such as relatively costly diesel and HFO.[6]

Of the six GUs studied in this chapter, half are considered small (AES Levant, AES PSC, and QEPCO); one, medium (AAEPC); and two, big (CEGCO and SEPCO). All are private except for SEPCO, and all rely on thermal production. CEGCO, AAEPC, and SEPCO have the worst ratio of employees to capacity, which could be an indication of overstaffing. Because Jordan is one of the only economies in the region not to have subsidies for fuel, its OPEX values are much higher than those observed in other economies in the region. This results in a high share of fuel costs, ranging from 95 percent to 98 percent of OPEX among GUs. Although CEGCO and SEPCO appear to almost fully recover their costs, QEPCO and AES PSC do not (and we did not have data for AES Levant). The debt-to-equity ratio in Jordan is mostly similar across GUs and in some cases lower than the MENA median. The GUs are all profitable, with high ROE and ROA values. In particular, the IPPs enjoy an ROE at least three times higher than the MENA median value of 5 percent. This is mainly a result of the provisions of the IPP contracts under which they operate and the PPAs with NEPCO that ensure that the IPPs can actually make a profit. Another reason is that NEPCO absorbs the fuel price fluctuation risk from GUs.

The three Jordanian DUs are private. For the purposes of our study, two are considered of medium size (IDECO and JEPCO) and one as small (EDCO). IDECO appears to be the most efficient DU in terms of operational performance. The value of OPEX per employee across all DUs is very high, presumably because electricity purchase prices reflect the fact that the fuels used to produce it are not subsidized. The only utility that fully recovers OPEX from sales is IDECO, although the values for EDCO and JEPCO are not far from cost-recovery. All Jordanian DUs have high debt-to-equity ratios, suggesting a high level of financing through debt. The three distribution utilities are notably profitable, as shown by the ROA and ROE indicators.

Jordan's main challenge is to provide affordable energy to end users while ensuring the profitability of mostly private generators and distributors in the wake of a significant energy transition. On the generation side, disruptions in the supply of natural gas have spiked interest in alternative sources of fuel, particularly renewable energy, which is high on the government's agenda since the adoption of the 2012 Renewable Energy Law. Other options include the potential use of vast oil shale resources. On the supply side, the main issue is to strike the right balance so as to (a) reduce dependence on subsidies while minimizing the impact of tariff reforms on the poorest consumers on the one hand and (b) reduce the fiscal deficits of utilities operating in an already challenging environment on the other.

Last but not least comes the issue of data collection and quality. Only part of the quantitative evidence on performance provided in this chapter is available online, and collecting the rest required numerous exchanges with the regulator and utilities to check the validity of figures and establish a common understanding of the factors behind specific values. Even after these efforts, some values were left aside because they did not appear to be reasonable. The exercise of carrying out periodic performance assessments with the direct involvement of both GUs and DUs should be a central task of the regulator. Jordanian utilities are generally well equipped to collect reliable information. All utilities report having implemented supervisory control and data acquisition (SCADA), and adopted international accounting standards (IAS). But most GUs have not yet implemented a cost accounting system, so this is a pending task.

Notes

1. Data for non-MENA generation utilities were insufficient to establish meaningful comparisons.
2. The total installed capacity in 2014 was 1,687 MW with available capacity of 1,267 MW whereas the installed capacity in 2009 was 1,747 MW with available capacity of 1,599 MW.
3. The indicators energy sales/total OPEX, energy sales/total costs, and accounts receivable are not applicable to generation utilities based on how these indicators were categorized for the purpose of the MENA Electricity Database. However, for comparative purposes, their values are presented and discussed in this chapter but not in previous chapters of this book.
4. The cost of fuel was estimated for the Jordanian generation utilities based upon the average cost of fuel per kilowatt-hour from the regulator and the amount of kilowatt-hours generated by each utility in 2013. This cost of fuel was then added to the operating costs to obtain the total OPEX as per the definition used in this study.
5. Figures from MED rounded to the nearest thousand.

6. After 2013, the Jordanian government started buying liquefied gas in the Port of Aqaba and transporting it to generation plants as a solution to this problem, which has had positive effects on the sector.

References

EMRC (Energy and Minerals Regulatory Commission). 2013. *Annual Report 2013*. Amman, Jordan: EMRC.

NEPCO (National Electric Power company). 2012. Annual Report 2012. Jordan: NEPCO.

NEPCO (National Electric Power company). 2013. *Annual Report 2013*. Jordan: NEPCO.

World Bank. 2017. MENA Electricity Database. World Bank, Washington, DC.

CHAPTER 8

Benefits and Challenges of Multiservice Providers: The Case of Morocco

The Ministry of Energy, Mines, and Sustainable Development (MEMDD) is in charge of energy sector policy in Morocco. Office National de l'Electricité et de l'Eau Potable (ONEE),[1] the national operator in charge of generation, transmission, and distribution in large parts of the country, is under the control of MEMDD but is under the financial supervision of the Ministry of Economy and Finance. With respect to power distribution, the Ministry of Interior oversees public municipal operators (*régies autonomes*) as well as private distributors (*sociétés délégataires*), which are elected by municipalities that grant public services concessions to the private sector. The setting and regulation of sector tariffs is the responsibility of an ad hoc interministerial committee (Commission des Prix) chaired by the Ministry of Governance and General Affairs (figure 8.1).

As of 2013, Morocco did not have an independent regulator, although the establishment of one had been under consideration for some time.[2] The electricity market was structured as a single-buyer model, in which ONEE acts as the sole buyer and supplier of bulk power.[3] ONEE supplies power from its own generation plants, purchases it from licensed independent power producers (IPPs) or through its international interconnections, and sells it to other distribution utilities and large industrial clients and through its own distribution grid.

Electricity Generation

ONEE is a state-owned vertically integrated utility (VIU), covering the generation (4,500 megawatts [MW]), transmission, and distribution of electricity. It has a monopoly on transmission operations and is the sole power supplier to distribution utilities. ONEE is also in charge of water and sanitation service delivery in large parts of the country. Whereas it is still the largest power producer of installed generation capacity (more than 63 percent), most electricity has been generated by IPPs since a 1994 amendment to the law governing ONEE's

Figure 8.1 Electricity Sector Organization, Morocco

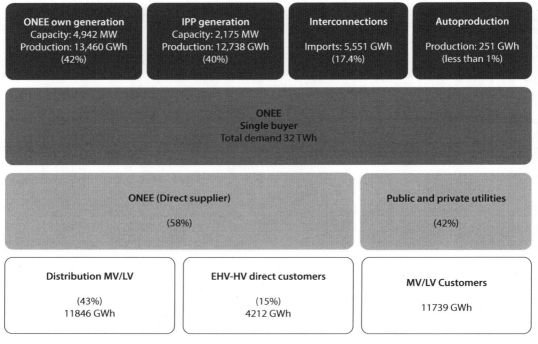

Source: Amegroud 2015.
Note: EHV = extra high voltage; GWh = gigawatt-hours; HV= high voltage; LV = low voltage; MV = medium voltage; MW = megawatts;
TWh = terawatt-hours.

activities opened up generation to private operators.[4] The remaining amount of electricity is imported from Spain and Algeria (5.5 terawatt-hours [TWh] in 2013) (Kharbat 2014). By the end of 2013, the total installed power capacity in Morocco was reported to be 7,994 MW, of which hydropower and renewable energy (wind and solar) represented 32 percent, including hydro pumped storage (see table 8.1).

Figure 8.2 illustrates the share of fuel types in generation. Thermal generation facilities are mostly used to produce electricity, with coal being the predominant fuel. Coal contributes significantly to the energy mix, with 31 percent of installed capacity; fossil-fuel-based power generation takes the lion's share, at 68 percent of total installed capacity. The contribution of hydropower is 22 percent (including hydro pumped storage), accounting for the largest share of renewable energy generation. Morocco's wind power production is the largest in the Middle East and North Africa (MENA) region, accounting for 10 percent of the nation's power generation capacity.

Four IPPs had a total installed capacity of 3,086 MW and supplied 52 percent of electricity in 2013.[5] The largest of these by far is the Jorf Lasfar Energy Company (JLEC), owned and operated by the Abu Dhabi National Energy Company PJSC (TAQA). JLEC runs the largest coal-fired power plant in MENA.

Table 8.1 Generation Mix, Morocco, 2013

Generation type (MW)	Amount
Hydropower generation	1,306
Pumped hydroelectric energy storage (PHES)	464
Steam power generation	3,145
Coal power generation (2,545 MW)	
HFO power generation (600 MW)	
Gas turbines power generation	1,230
Combined cycle power generation	850
Diesel power generation	202
Total thermal power generation	**5,427**
New and renewable energy (wind)	797
Total ONEE installed capacity (MW)	**7,994**
Total electricity generated (GWh)[a]	**28,081,540**

Source: ONEE 2014.
Note: GWh = gigawatt-hours; HFO = heavy fuel oil; MW = megawatts; ONEE = Office National de l'Electricité et de l'Eau Potable.
a. Does not account for imports from Spain.

Figure 8.2 Generated Electricity in Morocco, by Technology Share, 2013
Percent

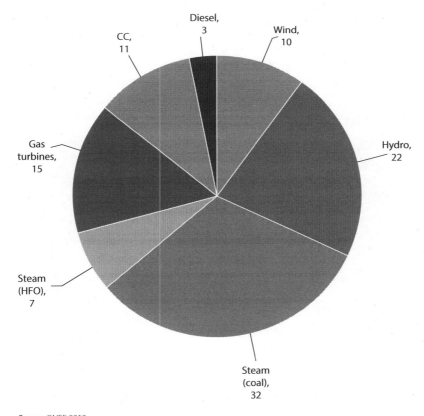

Source: ONEE 2013.
Note: CC = combined cycle; HFO = heavy fuel oil.

Electricity Transmission

Transmission activity is carried out by ONEE. The state-owned monopoly is responsible for managing, operating, and maintaining the electric transmission grid and interconnections with neighboring economies. The Moroccan power system is interconnected with the Spanish and Algerian grids. Table 8.2 provides some figures on electricity transmission by ONEE.

Electricity Distribution

While ONEE is in charge of power distribution in most of Morocco's cities and regions, there are 11 other electricity distribution entities, 7 public municipal utilities, and 4 private concession holders. With its 5.2 million clients, ONEE serves the largest number of consumers by far.

The second-largest distribution utility is Lyonnaise des Eaux de Casablanca (LYDEC), which delivers electricity in Casablanca and Mohammedia to 0.9 million consumers. REDAL, which covers distribution in Rabat and Sale, has 0.5 million customers (MENA Electricity Database [MED]).[6] The seven municipal public distributors, or *régies autonomes de distribution*, are RAEEF, in Fès; Régie Autonome de Distribution d'Eau d'Électricité et d'Assainissement liquide de la province de Kenitra (RAK), in Kenitra; RADEEL, in Larache; Régie Autonome de Distribution d'Eau et d'Électricité de Meknès (RADEM), in Meknès; Régie Autonome de Distribution d'Eau et d'Électricité de Marrakech (RADEEMA), in Marrakech; Régie Autonome de Distribution d'Eau, d'Électricité et d'Assainissement liquide des Provinces d'El Jadida et de Sidi Bennour (RADEEJ), in El Jadida; and Régie Autonome Intercommunale de Distribution d'Eau et d'Électricité de Safi (RADEES), in Safi.

All these utilities are multiservice operators offering water and sanitation services as well. The private distributors manage concessions in Casablanca-Mohammedia (LYDEC, a privately owned utility with Engie as the main shareholder), in Tangier and Tetouan (Amendis, part of the French utility Veolia), and in Rabat-Sale (Redal, part of Veolia). Table 8.3 includes basic data on the distribution network and numbers of customers.

Table 8.2 Electricity Transmission Data, Morocco, 2013

Transmission	Amount
Total transmission lines and cables (150 kV, 225 kV, 400 kV) km	22,995
High voltage (400 kV, 225 kV, and 60 kV/60 kV, 22 kV) substation capacity MVA	26,072

Sources: ONEE 2013 and MED.
Note: km = kilometers; kV = kilovolts; MED = MENA Electricity Database; MENA = Middle East and North Africa; MVA = megavolt ampere; ONEE = Office National de l'Électricité et de l'Eau Potable.

Figure 8.3 provides some insights on the heterogeneity of ONEE's client base. For 2013, it shows the breakdown of energy sales per customer usage. Whereas 50 percent of sales were to distribution utilities, the residential sector accounted for 20 percent, the industrial sector for 12 percent, and the agricultural sector for 7 percent.

Table 8.3 Electricity Distribution Data, Morocco, 2013

Transmission	Amount
Distribution transmission lines, medium voltage and low voltage (km)	243,568
Distribution substation capacity (MVA)	6,360
Customers (millions)	
Non-ONEE	2.9
ONEE	4.9
Total	7.8

Source: ONEE 2013.
Note: km = kilometers; MVA = megavolt ampere; ONEE = Office National de l'Électricité et de l'Eau Potable.

Figure 8.3 Share of Volume of Energy Distributed, by Sector, Morocco, 2013
Percent

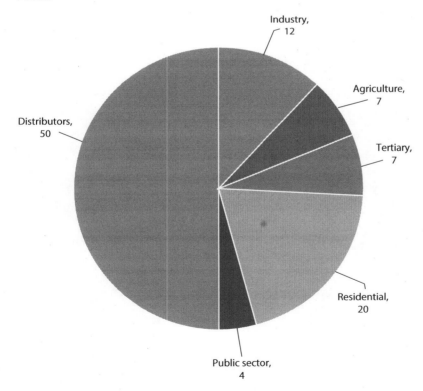

Source: ONEE 2013.

Comparison of Moroccan Generation Utilities

Although the sample used for this study did not include the four Moroccan IPPs, this chapter provides information on one of them, JLEC.[7] ONEE is also included in this section because it generates 42 percent of Morocco's electricity consumption. However, the financial and technical indicators related to its generation activity should be considered with caution, because it is difficult to separate ONEE's generation activity from the consolidated operating results (among other things, the utility provides water and sanitation services as well). Finally, and given the limited amount of data available for Moroccan generation utilities (GUs), we also look at an Egyptian GU (Upper Egypt Electricity Production Company [UEEPC]) as a comparator, in addition to the usual MENA median.

Table 8.4 compares ONEE, JLEC, UEEPC, and the MENA median across several indicators.

The Moroccan private GU has a capacity factor above the MENA median and similar to that of UEEPC. This could be a consequence of the power purchase agreement (PPA), whereby electricity purchase is guaranteed by

Table 8.4 Comparing the Performance of Moroccan Generators across Indicators and against Egypt's Upper Egypt Production Company and the MENA Median, 2013[8]

Category	Indicator name	Unit	ONEE	JLEC	Upper Egypt	Median MENA
General	Installed capacity	GW	4.9	2.0	3.4	—
	Net generation	TWh	13.0	13.5	21.0	—
	Employment	Employees	8,796	482	3,200	—
Technical and Operational	Capacity factor	%	31	75	70	58
	Availability factor	%	Between 75 and 80[a]	91	—	93
	OPEX/employee	$, thousands	284[b]	1,205	179	297
Financial (Cost structure)	Share of cost of fuel, lubricant in total OPEX	%		38.0	94.5	93.0	75.0
	Share of labor cost in total OPEX	%	10	5	5	12
Financial (Cost recovery)[c]	Energy sales/total OPEX	%	118	153	—	109
	Energy sales/total costs	%	87	129	—	107
Financial (Balance sheet)	Accounts receivable	Days	159[b]	45	—	40
	Debt/equity	%	—	277	1,270	357
	Current assets/current liabilities	%	63[b]	247	56	95
Financial (Profitability)	Return on assets	%	−4.4[b]	4.35	0.35	3.0
	Return on equity	%	—	17.3	3.0	7.0

Source: World Bank calculations.

Note: GW = gigawatts; JLEC = Jorf Lasfar Energy Company; MENA = Middle East and North Africa; ONEE = Office National de l'Électricité et de l'Eau Potable; OPEX = operating expenses; TWh = terawatt-hours; — = not available.

a. Average of all the utilities under ONEE is between 75 percent and 80 percent, depending upon the year.

b. Denotes values for ONEE, which are not disaggregated to the level of electricity generation.

c. The values of MENA medians above 100 percent are mainly driven by Omani generation utilities (12 out of 23 used in this study), which have high median values, at 109 percent and 107 percent, respectively, for the two cost-recovery indicators.

ONEE, and the utility is encouraged to maximize use of its generation capacity. The capacity factors varied between 31 percent in the case of ONEE's power generation activity and 75 percent for IPPs such as JLEC and Energie Electrique Tahaddart (EET).[9] These numbers show that private power production mainly serves the base load, and ONEE's facilities operate as load followers and for peaking.

The availability factor indicates the amount of time a power plant is available to generate electricity. This indicator varies greatly depending on type of fuel, plant design, and operations. It does not provide any indication of a plant's conversion performance or utilization rates. Morocco's private power producers generally boast high availability factors: 91 percent for JLEC and 93.4 percent in case of EET.[10] The availability of ONEE's thermal power plants range from an average of 53 percent for steam plants fueled by heavy fuel oil (HFO) to 98 percent for hydropower plants. The availability gaps observed in ONEE's generation facilities can mainly be explained by the existence of repair and maintenance issues. According to ONEE, its average availability factor ranges between 75 percent and 80 percent.

There are significant differences in terms of technical performance between ONEE and the private GUs: for example, JLEC outperforms ONEE's coal generation plants in terms of the heat rate. At the low end, one such coal plant, Jerada, requires almost twice the quantity of coal to produce the same amount of power as JLEC.[11]

JLEC's cost performance is strong and reflects its high availability and capacity factors. For GUs, operating expenses (OPEX) are associated with operating the power plant and generating electricity (fuel costs, maintenance, and administration). OPEX per employee values are mainly affected by the overall heat rate of the generating facility and its capacity factor, and also by its human resources management policy and the degree of reliance on outsourcing and subcontracting. Even so, OPEX per employee for JLEC is relatively high ($1.2 million).

The share of energy purchases and cost of fuel, lubricant, gas, and coal in total OPEX was 94.5 percent for JLEC. This is explained by the relatively high price of coal ($85 per ton)[12] and the high capacity factor of the plant. Though an accurate estimation of ONEE's generation activity OPEX is not available, we can say that JLEC's share of energy purchases, while still dominant, is much smaller because of ONEE's (a) aging generation facilities and hence high maintenance expenses, (b) larger number of employees and therefore relatively higher wage bill, and (c) relatively lower fuel bill as a result of a lower capacity factor.

Table 8.4 shows that JLEC exhibits strong performance indicators and a very healthy financial profile compared to that of Upper Egypt. JLEC's performance is to be viewed in light of its status as an IPP operating under a government-backed PPA, while Upper Egypt is a publicly owned utility.

JLEC's cost-recovery rate is comfortably high (153 percent for recovery of OPEX from sales), as would be expected from an IPP. This level reflects the strong profitability of the business. Morocco's private power producers enjoy attractive contractual arrangements, and PPAs are designed to pass most market and institutional risks to ONEE. Cost-recovery indicators for this category of GUs are strong.

The cost-recovery performance of ONEE's generating activity is difficult to assess without detailed analysis of the overall costs of the VIU and its sales. This is further complicated by cross-subsidies between different types of clients and a complex tariff structure. ONEE's electricity sales alone are not sufficient to cover total costs (that include depreciation and interest rates), as shown in the energy sales to total costs indicator, with a value of 87 percent in 2013.

The average number of days for receivables from sales is 45 days for JLEC, which is higher than the regional median for GUs, at 40 days, while the number of days for receivables from sales is 159 days for ONEE.

The debt-to-equity ratio for JLEC is 277 percent, which is relatively high but still much lower than the Egyptian GUs (1,270 percent for UEEPC). This is the typical ratio of an IPP, which often resorts to project financing to fund PPA-backed power generation infrastructures. In the case of the Arab Republic of Egypt, finance would be obtained in large part from the national budget in the form of direct subsidies allocated to the utility.

At 17.3 percent, JLEC's return on equity (ROE) was high, while its return on assets (ROA) was 4.35 percent. This compares positively with profitability performance indicators of GUs in the region, showing that the utility is highly profitable. ONEE, on the other hand, had a negative ROA in 2013 (−4.40 percent).

To conclude, the impact of an imbalanced pricing structure has pushed ONEE to reduce or delay investments in maintenance and performance improvements, thereby increasing its focus on fulfilling its obligations under signed PPAs and electricity imports from Spain. Furthermore, the extensive use of IPPs with government-backed PPAs since 1994 has resulted in a situation where performing assets are owned by new entrant private investors while a large proportion of risks are passed to ONEE (for example, fuel price, exchange rate). Distribution is also organized in such a way as to shield large distributors from market risks and the impact of volatile power generation costs.

Comparison of Moroccan Distribution Utilities

Table 8.5 presents indicators for the 11 distribution utilities in Morocco, as well as the MENA median values.

Load factor values were obtained for all utilities; they ranged from a low of 45 percent for AMENDIS Tetouan to a high of 64 percent for RADEEJ. High load factors are representative of the load profile and indicate that the ratio of peak demand to average demand is relatively low. This in turn indicates that the industry's stable consumption comprises a significant share of total demand. RADEEJ, LYDEC, AMENDIS Tanger, RAK, and REDAL had load factors higher than 56 percent, which is consistent with the industrial role across MENA (where the median value is 60 percent). Distribution losses in Morocco are lower

Table 8.5 Comparing the Performance of Moroccan Distributors across Indicators and against the MENA Median, 2013

Category	Indicator name	Unit	AMENDIS Tanger	AMENDIS Tetouan	LYDEC	RADEEL	REDAL	RAK	RADEEMA	RADEM	RADEEJ	RADEEF	RADEES	MENA Median
Technical and operational	Load factor	%	56	45*	58*	53*	56*	57*	51	48*	64	46*	52*	60
	Distribution losses	%	10	11	7	8	8	8	5	7	4	—	3	10
	OPEX/employee[a]	$ thousands	321	151	527	203	642	248	287	254	190	186	200	188
	OPEX/connection	$	508	346	836	361	644	412	410	309	396	318	339	346
	OPEX/kWh sold	$	0.12	0.15	0.20	0.12	0.17	0.12	0.10	0.11	0.10	0.11	0.13	0.1
	OPEX/km	$ thousands	36	31	96	19	49	20	32	21	21	37	32	19.6
Commercial (Consumption and billing)	Total billing/ connection	$ thousands	473	299	520	—	442	306	466	301	436	312	302	299
Financial (Cost structure)	Share of labor cost in total OPEX	%	—	—	12	—	14	—	8	—	12	—	—	12
Financial (Cost-recovery)	Energy sales/OPEX	%	—	—	100	86	103	94	130	97	136	98	89	93
	Energy sales/total costs	%	—	—	89	—	92	—	—	—	119	—	—	88
Financial (Balance sheet)	Accounts receivable	Days	—	—	76	—	121	—	205	—	106	—	—	121
	Debt/equity	%	—	—	279	—	—	—	41	—	66	—	—	523
	Collection rate	%	—	—	—	—	—	—	—	—	—	—	—	93
	Current assets/ current liabilities	%	—	—	72	—	92	—	—	—	64	—	—	85
Financial (Profitability)	Return on assets	%	3	-1	—	6	2	—	—	21	—	—	14	3
	Return on equity	%	3	-2	18	7	10	—	—	22	—	—	16	7

Source: MENA Electricity Database except when marked with a "*" in which case obtained directly from utilities.

Note: For LYDEC and REDAL, OPEX values are consolidated figures that include activities other than power distribution. OPEX = operating expenses; km = kilometers; kWh = kilowatt-hours; MENA = Middle East and North Africa; — = not available.

a. Values reflect the estimated number of employees needed to support a utility's electricity activities.

than or equal to the MENA median of 10 percent, with the exception of AMENDIS Tetouan at 11 percent.

Among municipal distributors, OPEX per employee varies from a minimum of $151,000 per employee for AMENDIS Tetouan, to a maximum of $642,000 per employee for REDAL. The number of employees provided by utilities was not disaggregated by function, and an estimate of those focused on electricity was used for the purposes of this study.[13]

The share of labor costs in total OPEX was low for the utilities that reported values, not exceeding 14 percent (REDAL). At 8 percent, RADEEMA reported the lowest value, while LYDEC and RADEEJ reported values of about 12 percent. This suggests that the OPEX are mainly made up of other costs such as the purchase costs of electricity from ONEE.

Disregarding LYDEC and REDAL, the OPEX per connection was highest for AMENDIS Tanger ($508 per connection) and lowest for RADEM ($309 per connection). The separation of electricity and water within these municipal utilities is not as clear as within ONEE, and electricity services are often used to cross-subsidize the water services (as well as the heavy investments required in sanitation-related activities). Hence, it is common that tariffs are not related solely to energy consumption (for example, meter renting and maintenance, technical interventions, specific studies, or opening and closing accounts) and are fixed by each operator following municipal agreements. These fees can represent a substantial amount of the operator's total revenues and are sometimes used by the operators to compensate for low national tariffs.

RADEEJ has the lowest OPEX per kilowatt-hour sold ($0.10 per kilowatt-hours [kWh]). RADEEJ operates in a region in which the weight of industrial activities (medium-voltage clients) is significant, therefore explaining the low cost of maintenance per energy sales.

With regards to the OPEX per kilometer (km) among the 11 distribution utilities, it costs the most to maintain and operate 1 km of the existing distribution network for LYDEC ($96,000 per km) while these costs are the lowest in the case of RADEEL ($19,190 per km).

Considering the limited data available, it appears that as shown in table 8.5, LYDEC, REDAL, RADEEMA, and RADEEJ positively recover their operating costs from energy sales, whereas all the other utilities show values that are close but still below 100 percent.

In terms of ROA, AMENDIS Tetouan showed negative values, while all Moroccan utilities showed low positive values, with the lowest shown by REDAL (2 percent) and the highest by RADEM (21 percent). The same trend for ROA is reflected in ROE, whereby AMENDIS Tetouan once again is the only utility with a negative value (−2 percent). In Morocco, RADEM reported a ROE of 22 percent, which is more than three times the MENA median.

Although ONEE engages itself in distribution activities, it was excluded from this analysis because it does not publish usable commercial and financial data on its distribution business. Some of the data published or provided by the two largest private distribution utilities, LYDEC and REDAL, are consolidated across all business activities (OPEX, head count, labor cost). The data used in the case of public municipal utilities (*régies*) were a combination of information directly collected from the utilities and data from the 2014 report on municipal public services, prepared by the Direction des Régies et Services Concédés.

The discrepancies between Moroccan power distribution utilities mainly reflect the specific features of each utility. These include the following:

- *Type of customers.* An important share of medium-voltage customers tends to push the load factor higher, as well as cost-recovery ratios, while keeping employment needs low (RADEEJ).
- *Economic activity.* A thriving economy reflected by relatively high standards of living tends to have a positive impact on recovery and profitability indicators. Tariffs charged to households with high electricity consumption are generally more profitable (LYDEC, REDAL).
- *Geography.* Operational costs and investments are generally higher for utilities operating in extended geographical areas, covering scattered clients (RADEEL).
- *Local climate.* Utilities operating in regions in the central part of the country are generally faced with higher operating and investment costs (RADEEMA, RADEEF).

Evolution of Morocco's Electricity Sector since 2014

The electricity sector in Morocco has gone through a number of changes since 2014 that are worth presenting, given that the analysis of this chapter is based on 2013 data.

To alleviate ONEE's poor financial state while simultaneously pushing the utility to improve its operational performance, in 2014 the government of Morocco and ONEE signed a framework contract (contrat programme) for the period 2014–2017. The goal of this financial restructuring plan was to help ONEE overcome a long-running precarious state of affairs. It focuses on tariff rate revisions, supplemented with an increase in capital and active help to collect former receivables from municipal utilities, public administrations, and municipalities. The plan also included a lump-sum payment to ONEE as a one-time flat subsidy for fuel oil used in electricity production to pave the way for a complete phase-out of all forms of oil subsidies. As a result, the net producer has gone from a deficit of about $285 million at the end of 2013 to a surplus of about $80 million at the end of 2016.

In 2015 self-generation was further opened with two measures. First, the power sector's legal framework for self-generation (above 300 MW) was further extended with the suspension of restrictions on capacity, type, and site of generation, provided that the total installed capacity be above 300 MW.

The new amendment recognizes the importance of providing more flexibility in terms of energy management to large industrial consumers such as in the mining sector. Second, for small self-generators, the scope of the renewable energy law was further widened to distribution grids. Private developers of renewable energy were allowed to connect their projects to the medium voltage grid and were given access, albeit with some restrictions, to the end users. This step will be taken further to allow households and small businesses (and all other low-voltage clients) to install on-grid renewable distributed generation equipment (such as rooftop solar kits), with the amendment of Law 13-09 in August 2015.

In 2016, Morocco adopted a law introducing an independent energy regulator (Agence Nationale de Régulation de l'Energie—ANRE) and detailing its functions, missions, and organization. The role conferred on this new authority will be confined to policing the power generation regime introduced under the renewable energy law. The law also paves the way toward the separation of ownership and operations of grids with generation and commercial activities. This cautious approach to the introduction of an independent regulation authority is considered to be more realistic than pursuing a body with wider prerogatives in an environment where the politics do not necessarily favor an active independent authority.

Finally, six years after the creation of the Moroccan Agency for Solar Energy (MASEN), in 2016 the Moroccan government decided to extend the agency's prerogatives to include the development and operation of all types of renewable energy facilities. The agency was renamed the Moroccan Agency for Sustainable Energy, and ONEE was required by law to transfer all its renewable energy assets to the new entity. This measure aims at emphasizing the role of renewable energy in future sector development and fast-tracking the implementation of the country's targets in terms of the overall share of renewables in power production.

Conclusion

This chapter analyzed the performance of Moroccan electricity utilities in 2013. A singular characteristic of Morocco's power sector is that most of its electricity service providers also provide water and sanitation services. The rationale was to use the relatively comfortable margins from electricity sales to subsidize water sales and finance investments in sanitation infrastructure networks and waste water treatment.

ONEE is considered to be a big VIU in this study, with more than 2 million connections. Beyond being the single buyer in Morocco, it produces more than 40 percent of the electricity and distributes almost 60 percent of it. In addition to ONEE, in the generation segment we have included the biggest of the four Moroccan IPPs, JLEC. We find significant differences in terms of technical performance between ONEE and the private GUs: for example, JLEC's heat rate outperforms ONEE's coal generation plants. JLEC's cost performance is strong and reflects its high availability and capacity factors. In addition, it exhibits a very healthy financial profile compared to that of Upper Egypt. JLEC's performance levels are to be viewed in light of its status as an IPP operating under a government-backed PPA, while Upper Egypt is a publicly owned utility. JLEC's cost-recovery rate is comfortably high (153 percent for the recovery of OPEX from sales), as would be expected from an IPP. This level reflects the strong profitability of the business. Morocco's private power producers enjoy attractive contractual arrangements, and PPAs are designed to pass most of the market and institutional risks to ONEE. Cost-recovery indicators for this category of GUs are strong. The cost-recovery performance of ONEE's generating activity is difficult to assess without detailed analysis of the entity's overall costs and sales. This is further complicated by cross-subsidies between different client types and by a complex tariff structure.

This study also analyzed 11 distribution utilities, of which 5 were medium and 6 were small as defined by this study. Distribution utilities in Morocco are also involved in water and sanitation activities. They all purchase electricity from ONEE. All the small distribution utilities are public (municipal distributors) except for AMENDIS Tetouan, which is private, whereas three of the five medium utilities are private. The distribution utilities perform fairly well at the technical level, with values on technical performance indicators close to the MENA medians. While labor costs of all distribution utilities make up 8 percent to 14 percent of OPEX, the ratio of OPEX to employees is higher than the MENA median for all except two utilities (RADEEF and AMENDIS Tetouan). All private utilities have strong profitability ratios, with the exception of AMENDIS Tetouan, which had a negative ROA and ROE. This can be explained by the fact that AMENDIS Tanger and Tetouan are actually a joint concession, and the private operator compensates Tetouan with the business in Tangier. Overall, data availability was the main obstacle to analyzing the financial performance of the municipal distributors.

Last but not least come the issues of data collection and data quality. Only part of the quantitative evidence on performance provided in this chapter was publicly available for ONEE and JLEC. Very little is publicly available on the other Moroccan electricity utilities.[14] Some values collected from utilities were left aside since they did not appear to be reasonable. There is space in Morocco to increase the performance information publicly available, which would also help the government carry out its regulatory tasks for operators. The issue of quality is of concern for public distribution utilities but not for private distribution utilities. The good news is that almost all Moroccan electricity utilities have implemented supervisory control and data acquisition (SCADA). Meanwhile, only private distribution utilities have adopted international accounting standards (IAS), and no public distribution utility reports data using cost accounting systems, while ONEE and private distribution utilities do use them.

Notes

1. Since 2012 and the adoption of law no. 40-09, the water and electricity activities of the former ONE (electricity) and ONEP (water) have been regrouped under one holding company, ONEE.

2. A 2015 law has since provided for the establishment of the ANRE (Agence Nationale de Régulation de l'Electricité), but its role is limited to the regulation of the renewable energy sector and associated transmission activities (tariffs and access conditions).

3. An incipient liberalized model has emerged under the framework of law no. 13-09. In 2014, these liberalized markets covered 2 percent of total electricity demand. In this framework, private large consumers have direct access to the transmission grid and can purchase part or all of their electricity from private renewable energy producers.

4. Decree No. 2-94-503 (September 1994).

5. None of these IPPs or generation utilities were included as part of the 67 utilities of this study due to lack of data availability. However, JLEC is included in this country case study.

6. See World Bank's MENA Electricity Database.

7. The reason being that Taqqa Morocco shares floated on the Casablanca stock exchange in 2013, and it was therefore required to publish its financial statements on a regular basis.

8. The indicators energy sales/total OPEX, energy sales/total costs, and accounts receivable are not applicable to generation utilities based on how these indicators were categorized for the purpose of the MENA Electricity Database. However, for comparative purposes, their values are presented and discussed in this chapter, but not in previous chapters of this book.

9. See Energie Electrique de Tahaddart website, http://eet.ma/decouvrir-notre-activite.

10. See EET's website (http://www.eet.ma). The 384 MW combined-cycle gas turbine power plant in Tahaddart is owned by EET. Shareholders include the Moroccan ONEE, the Spanish Endesa, and Siemens.

11. While JLEC has a heat rate of 2,195 kilocalories per kilowatt-hour, the values for the Jerada and Mohammedia coal plants—both pertaining to ONEE—are 3,850 and 2,534, respectively.

12. See the Intercontinental Exchange Futures Database, https://www.quandl.com/data/ICE/ATWK2013-Rotterdam-Coal-Futures-May-2013-ATWK2013.

13. It was estimated that electricity-related employees made up about one-third of all employees in Moroccan utilities (which carry out water, sanitation, and electricity activities). This was based upon actual figures for certain utilities (for example, RADEEMA, which has 350, 302, and 272 employees for a total of 921).

14. A yearly report on public distribution utilities prepared by the Ministry of Interior provides part of the information collected in this chapter.

References

Amegroud, T. 2015. "Morocco's Power Sector Transition: Achievements and Potential." Paper produced in the IAI-OCP Policy Center partnership, Istituto Affari Internazionali (IAI). http://www.iai.it/sites/default/files/iaiwp1505.pdf.

Kharbat, F. 2014. "Interconnection between Arab Countries." Presentation at the "Enabling Renewable Energy in the Electricity Systems" workshop, April 16, Tunis. Arab Union of Electricity. http://www.medelec.org/media/1066/khabat-2014-04-16.pdf.

LYDEC (Lyonnaise des Eaux de Casablanca). 2013. *Annual Activity Report, 2013*. Morocco: LYDEC.

———. 2014. *Annual Activity Report, 2014*. Morocco: LYDEC.

Ministère de l'Intérieur. 2014. *Direction des Régies et des Services Concédés—Activités*. Morocco: Ministère de l'Intérieur.

ONEE (Office National de l'Electricité et de l'Eau Potable). 2013. *Annual Activity Report 2013*. Morocco: ONEE.

———. 2014. *Annual Activity Report 2014*. Morocco: ONEE.

TAQA (Abu Dhabi National Energy Company, PJSC). 2013. *Annual Activity Reports 2013*. Abu Dhabi: TAQA.

———. 2014. *Annual Activity Reports 2014*. Abu Dhabi: TAQA.

CHAPTER 9

A Remarkably Sophisticated Power Market: The Case of Oman

In Oman, before 2004, the Ministry of Housing, Electricity, and Water (MHEW) was the sole body responsible for the purchase, transmission, distribution, and supply of electricity on the country's main interconnected system (MIS) and on its rural systems. Then in 2004, the Law for the Regulation and Privatization of the Electricity and Related Water Sector (the Sector Law), promulgated by Royal Decree 78.2004, significantly altered the way the country's electricity sector was organized, managed, and regulated. The electricity functions of the MHEW were transferred to nine newly established government-owned successor companies. The sector was thus vertically and horizontally unbundled.

All nine of these successor companies remain government owned except for Al-Rusail Power Company, which was privatized in 2007. The Electricity Holding Company SAOC (EHC), which was established at the same time as the successor companies and is 100 percent owned by the Ministry of Finance, held 99.99 percent of shares of the successor companies, while the Ministry of Finance directly held 0.01 percent of the shares.

Oman's power systems are not fully interconnected. Oman's MIS covers the northern part of the country. A smaller system known as the Dhofar Power Company (DPC) serves the Salalah region in the south. There is also a dedicated small system owned by Petroleum Development Oman (PDO)—the country's most important state-owned oil producer—with a capacity of 1,500 megawatts (MW). The transmission grids of MIS and DPC are both connected to the PDO system with very limited transfer capacity. Other areas are served by the Rural Areas Electricity Company (RAECO).

Since its establishment in 2005, the Authority for Electricity Regulation (AER) has undertaken numerous important projects to support the electricity sector in Oman. These have included reviewing and approving electricity and water-related bulk supply tariffs, as well as reporting on major developments in the electricity and water sectors.[1]

In 2013, the electricity sector in Oman was largely unbundled except for the rural system where RAECO is still a vertically integrated utility (VIU) responsible

Figure 9.1 Electricity Sector Organization, Oman

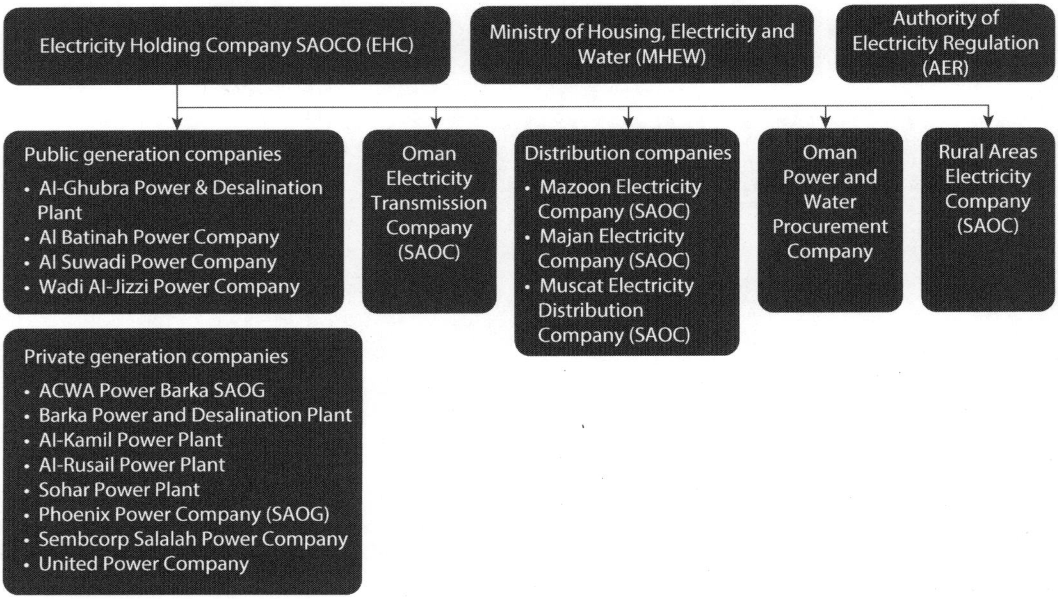

Source: World Bank.

for the generation, transmission, distribution, and supply of electricity to customers in its (mostly rural) concession area. Figure 9.1 shows the structure of the electricity sector in Oman in 2013.

The Oman Power and Water Procurement Company (OPWPC) is the single buyer of capacity and output from licensed production facilities and other entities, whereas the Oman Electricity Transmission Company (OETC) is a monopoly provider of transmission services to the MIS.

The Rusail Power Company, Wadi Al Jizzi Power Company (WAJPCO), and Al Ghubrah Power and Desalination Company (GPDCO) are electricity generators, while Mazoon Electricity Distribution Company (MZEC), Majan Electricity Company (MJEC), and Muscat Electricity Distribution Company (MEDC) each have monopoly rights to distribute and supply electricity within authorized areas stipulated in their respective licenses.

Electricity Generation

Electricity generation in Oman is covered by 12 generation utilities (GUs), two VIUs, and three distribution utilities (DUs). All the electricity produced is purchased by OPWPC. At the end of 2014, total installed capacity in Oman was 8,143 MW (table 9.1).[2] Around 88 percent of this capacity (7,191 MW) was on the MIS, 9 percent (718 MW) was on the Dhofar Power Company system, and the remaining 3 percent was on the rural power system. Ninety-eight percent of the fuel used to produce power (and desalinate water) for the MIS is natural gas supplied by the Ministry of Oil and Gas (the other 2 percent is oil).

Table 9.1 Generation Mix, Oman, 2013

Generation Type (MW)	Amount
Thermal power generation	8,143
Total installed capacity	8,143
Total generated energy (GWh)	28,343

Source: AER 2014.
Note: GWh = gigawatt-hours; MW = megawatts.

Electricity Transmission

Transmission of electricity in Oman is ensured by OETC, which was established as a monopoly provider of transmission services to the MIS.

In 2013, 2.7 percent of total net energy generated and the energy purchased was lost during transmission. Table 9.2 includes basic data for the transmission lines and substations of the transmission utility (TU).

Table 9.2 Electricity Transmission Data, Oman, 2013

Transmission	Amount
Total transmission lines and cables (230 kV, 150 kV, 132 kV) km	4,405
High voltage substation capacity (MVA)	20,700

Source: AUE 2013.
Note: km = kilometers; kV = kilovolts; MVA = megavolt ampere.

Electricity Distribution

Three DUs supply the MIS, namely MEDC, which in 2014 had the largest number of customers (261,480); MZEC (318,182 customers); and MJEC (174,592 customers) (AER 2014: 72). Table 9.3 includes some information on the distribution network, and the total number of customers in Oman.

There are significant differences between the DUs in the volumes of electricity supplied to different customers. Although more than 60 percent of the electricity MZEC supplies goes to residential customers, only 36.6 percent of MJEC's and 48 percent of MEDC's electricity supply goes to residential customers. In turn, more than 40 percent of the electricity that MJEC supplies goes to industrial customers, whereas only 6.3 percent of MEDC's and only 1.6 percent of MZEC's electricity supply goes to industrial customers (AER 2014, 73).

Figure 9.2 provides insights on the heterogeneity of the utilities' client base. It reports the percentage of energy sold in 2014 by customer type and use. The residential sector accounts for 75 percent of the customer base and also buys the largest share (at 35 percent of all energy distributed). The public sector accounts for 4 percent of the customer base, yet consumes 16 percent of the electricity distributed.

Shedding Light on Electricity Utilities in the Middle East and North Africa
http://dx.doi.org/10.1596/978-1-4648-1182-1

Table 9.3 Electricity Distribution Data, Oman, 2013

Distribution	Amount
Distribution lines length (km)	40,552
Distribution substation capacity (MVA)	10,299
Consumers (millions)	0.81

Sources: AUE 2013; OPWPC 2014.
Note: km = kilometers; MVA = megavolt ampere.

Figure 9.2 Share of Energy Distributed, by Consumer Sector, Oman, 2013
Percent

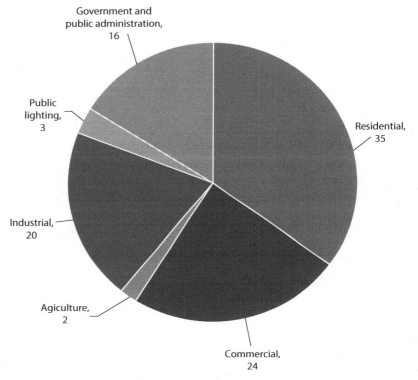

Source: MENA Electricity Sector Assesment Report.

Comparison of Generation Utilities in Oman

Oman's unbundled market allows the comparative assessment of generation and distribution. The partial indicators reported here are only a first step toward a full diagnostic since they need to be considered collectively and corrected for the specific supply and demand conditions faced by each individual utility. With this limitation in mind, these partial indicators compared the performance of the two utility types against one another, as well as with the Middle East and North Africa (MENA) median values.

As seen from the general indicators in table 9.4, data gaps are significant for the utilities in Oman. The comparison is therefore limited to utilities for which data were available for each indicator.

Table 9.4 Comparing the Performance of Oman's Generation Utilities across Indicators and against the MENA Median, 2013[3]

Category	Indicator name	Unit	APBS	ABPC	ASPC	GPDCO	AKPP	ARPP	BPDP	PPC[a]	SSPWC	SPP	UPC	WAJPCO	MENA median
General	Installed capacity	MW	456	744	744	499	285	687[b]	672	200	489[b]	585	270	245	—
	Net generation	GWh	2,372	2,655	2,083	2,514	1,672	3,459	3,029[b]	—	1,860[b]	3,538	1,194	474	—
	Employment	Employees	74	65	50	273	—	16	—	—	80	76	48	—	—
	Fuel mix	Gas	Gas	Gas	Gas	Gas	Gas	Gas	Gas	Gas	Gas	Gas	Gas	Gas	—
	OPEX	$ millions	—	44.0	41.0	83.0	37.0	80.0	33.0	0.3	49.0	72.0	14.0	19.0	—
Technical and operational	Capacity factor	%	59	41	32	58	67	—	—	—	—	69	51	—	58
	Availability factor	%	93	96	90	85	89	—	—	—	—	93	91	—	93
	OPEX/employee	$ thousands	—	676	816	304	—	—	—	—	615	942	289	—	297
Financial (Cost structure)	Share of cost of fuel, lubricant in total OPEX	%	—	59	61	75	78	77	52	—	51	68	—	47	75
	Share of labor cost in total OPEX	%	—	—	—	13	—	—	—	—	—	—	—	25	12
Financial (Cost-recovery)	Energy sales/total OPEX	%	—	256	276	127	146	110	314	—	246	173	214	108	109
	Energy sales/total costs	%	—	135	136	107	120	—	123	—	86	112	108	97	107
Financial (Balance sheet)	Accounts receivable[b]	Days	36	18	19	46	24	5	33	—	44	30	54	58	40
	Debt/equity	%	249	303	294	—	94	—	1,857	—	357	1,399	72	—	357
	Current assets/current liabilities	%	121	54	53	443	79	156	42	—	179	118	38	504	95
Financial (Profitability)	Return on assets	%	8	—	—	1	9	—	3	0.1	3	3	5	8	3
	Return on equity	%	24.0	—	—	0.2	15.0	—	—	—	13.0	—	7.0	2.0	7.0

Source: World Bank calculations.

Note: ABPC=Al Batinah Power Company; AKPP = Al-Kamil Power Plant; APBS = ACWA Power Barka; ARPP = Al-Rusail Power Plant; ASPC = Al Suwadi Power Company; BPDP = Barka Power and Desalination Plant; GPDCO = Al-Ghubra Power and Desalination Company; GWh = gigawatt-hours; MENA = Middle East and North Africa; MW = megawatts; OPEX = operating expenses; PPC= Phoenix Power Company; SPP = Sohar Power Plant; SSPWC = Sembcorp Slalah Power and Water Company; UPC = United Power Company; WAJPCO = Wadi Al-Jizzi Power Company; — = not available.

a. Commercial operation started on December 11, 2014.

b. Values obtained from annual reports and not present in the MENA Electricity Database.

All the GUs for which data were available have installed capacities above 200 MW. All utilities in Oman are either small or medium (except for Phoenix Power Company), and five of the GUs are publicly owned whereas the rest are private. Several generation plants in Oman are also involved in desalination and other water sector activities.

In terms of electricity generated, the lowest value recorded was for WAJPCO (474 gigawatt-hours, GWh), whereas the highest, shown in table 9.4, was for Sohar Power Plant (SPP) (3,538 GWh). The number of employees in Oman's GUs remains low. The utilities are thermal power plants, using natural gas.

The capacity factors vary between 32 percent and 69 percent. These values are comparable to the normal range observed for thermal units, although they are on the lower bound. Yet some utilities in Oman perform better than the MENA median of 58 percent, such as SPP, with the highest capacity factor among the observations in table 9.4.

GUs' availability factor was relatively high, ranging from 85 percent (GPDCO) to 96 percent (ABPC). This is because all electricity generation plants on the MIS use gas turbines to generate electricity. The availability factor—that is, the ratio of the in-service time period to the total year—depends on generation outages, whether due to failure or maintenance. It also depends on the availability of fuel, yet doesn't indicate whether the units are working at full or partial capacity.

The differences between certain indicator values can be explained by the fact that some generation plants are also involved in desalination activities. Most of these plants are private entities that have small staff numbers compared to other utilities in the region. Taking into consideration the generation technology, fuel types, staffing characteristics, and general site layout can help clarify the differences between the GUs. For example, some plants in Oman are gas turbine units used for peaking only. This is characteristic of a high-income country in the MENA region.

Operating expenses (OPEX) per number of employees is higher among GUs in Oman when compared to the MENA median. This is mainly driven by the low employment levels in Oman, as shown in table 9.4.

The cost of gas accounts for a significant part of the OPEX, representing as much as 78 percent (Al-Kamil Power Plant [AKPP]). In general, the share of fuel cost in total OPEX of utilities in Oman is similar to, or lower than, the MENA median of 75 percent.

All utilities recover their total OPEX from sales of energy. The Barka Power and Desalination Plant (BPDP) recovers total OPEX at 314 percent, which is the highest value among the utilities. Performance for this indicator is generally much higher among GUs in Oman than the MENA median of 109 percent. This could be a result of the attractive purchasing prices charged by the GUs, as per their power purchase agreements (PPAs) with the single buyer. A similar trend is observed for the recovery of total costs from the sales of energy (except for the

Sembcorp Salalah Power and Water Company [SSPWC], which has a cost-recovery of 86 percent).

With regard to the time lapse between accounts receivable and sales, values range from as low as 18 days (ABPC) to as high as 58 days (WAJPCO), which is not far from the MENA median of 40 days. These values allow the utilities to have a constant flow of cash at hand.

The debt-to-equity ratio varies from 94 percent as observed for AKPP to 1,857 percent for BPDP. A debt-to-equity ratio that is too low could imply that the utility has a large amount of cash on hand and is, therefore, not necessarily managing its equity in the most efficient way, whereas a very high ratio would imply that the utility has a high level of debt and depends on debt to finance its projects and operations. Ideally, a ratio value oscillating around 100 percent would show that the utility is capable of managing its equity while at the same time using debt as a strategic financing tool.

For half the utilities, the ratio of current assets to current liabilities was 100 percent or more. In the other half, ABPC, the Al Suwadi Power Company (ASPC), AKPP, BPDP, and the United Power Company (UPC) had values in the range of 38 percent to 79 percent. For these utilities, current liabilities are not liquid.

The return on assets (ROA) for generators in 2013 ranged from 0.1 percent (Phoenix Power Company [PPC]) to 9 percent (AKPP), whereas the return on equity (ROE) ranged from 0.2 percent (GPDCO) to 24 percent (ACWA Power Barka [APBS]). The ROA for PPC, as shown in table 9.4, is low mainly because it only became operational in 2014. The profitability of generators in Oman is derived from their availability and reliability. Changes in the demand and supply landscape do not affect profits because a pass-through cost exists for the GUs.

Comparison of Distribution Utilities in Oman

Table 9.5 compares the three DUs MJEC, MZEC, and MEDC. The MENA median values are also included for comparison.

The load factor of the three DUs ranged from 44 percent to 71 percent, largely reflecting differences in their customers base. Commercial and industrial customers tend to have a higher load factor than residential customers. Because commercial and industrial customers account for 56 percent of MJEC's total supply, compared to 36 percent in the MIS overall, MJEC has a relatively higher load factor. Distribution losses, on the other hand, are relatively high among the Omani DUs, ranging from 9 percent to 13 percent.

OPEX per employee is of the same order of magnitude for MZEC and MEDC ($226,672 and $174,580, respectively). Values of this indicator are higher than the MENA median ($188,000). Labor costs' share of OPEX in Omani DUs remains low (5 percent for MEDC and 6 percent for MJEC), showing that the OPEX is most probably made up of other costs such as electricity purchase costs.

Table 9.5 Comparing the Performance of Oman's Distributors across Indicators and against the MENA Median, 2013

Category	Indicator name	Unit	MJEC	MZEC	MEDC	MENA Median
Technical and operational	Load factor	%	71	44	55	60
	Distribution losses	%	13	11	9	10
	OPEX/employee	$ thousands/ employee	227	175	—	188
	OPEX/connection	$/connection	—	1,150	1,698	346
	OPEX/kWh sold	$/kWh	0.05	—	—	0.1
	OPEX/km	$, thousands/km	—	14	42	19.6
Financial (Cost structure)	Share of labor cost in total OPEX	%	6	—	5	12
Financial (Cost-recovery)	Energy sales/OPEX	%	69	61	80	93
Financial (Balance sheet)	Accounts receivable	Days	119	110	122	121
	Debt/equity	%	109	148	147	523
	Collection rate[a]	%	79	77	74	93
	Current assets/ current liabilities	%	43	18	46	85
Financial (Profitability)	Return on assets	%	8	6	8	3
	Return on equity	%	14	14	16	7

Source: World Bank calculations.

Note: km = kilometers; kWh = kilowatt-hours; MEDC = Muscat Electricity Distribution Company; MJEC = Majan Electricity Company; MZEC = Mazoon Electricity Distribution Company; MENA = Middle East and North Africa; OPEX = operating expenses; — = not available.

a. Values obtained from calculations in appendix C and not present in the MENA Electricity Database.

Accounts are received within 119 days for MJEC, 110 days for MZEC, and 122 days for MEDC. With regard to cost-recovery performance indicators, values of energy sales to total OPEX were available for all three DUs, with MEDC having the highest total OPEX recovery from sales, at 80 percent. However, no values were available for total cost recovery from sales of electricity and for total billing per connection.

Current assets to current liabilities ratio for the three utilities were 43 percent, 18 percent, and 46 percent, respectively. MZEC has the lowest liquidity ratio. All three DUs have low values, which suggests that they are unable to repay their current liabilities (as could be the case with the lower value observed for MZEC) or that they are managing their current assets in a strategic manner (as could be the case for MEDC, because the ratio is closer to 50 percent).

Finally, in terms of profitability all three DUs showed positive results. MEDC and MJEC have similar ROA—8 percent each. MEDC also has the highest ROE, at 16 percent, followed by MJEC and MZEC, at 14 percent each.

Evolution of Oman's Electricity Sector since 2014

The electricity sector in Oman has gone through a number of changes since 2014 that are worth mentioning, given that the analysis of this chapter is based on 2013 data. The system peak demand in MIS grew by 9 percent annually between 2009 and 2014, reaching 5,122 MW in 2014, and was forecast to grow at about the same rate to 9,530 MW in 2021. Energy requirements were expected to grow from 25 terawatt-hours (TWh) to 47.1 TWh during the seven-year period 2015–21. Fuel use was expected to increase by 4 percent per year. Natural gas consumption as fuel was expected to increase by 4 percent a year. There are planned projects to interconnect the MIS and DPC grids and to extend the grid to some of the rural areas gradually. As Oman is dependent on hydrocarbon exports, which account for around two-thirds of total export earnings, it remains vulnerable to fluctuations in oil prices. Oil and gas revenue fell by 40 percent in 2015 due to lower oil prices, despite higher output (natural gas production rose by 5 percent in 2015).

In particular, significant energy subsidy reforms have taken place over the past few years. Since January 2015, Oman doubled gas tariffs for industrial producers and the power industry to $3.0/million British thermal units. In the absence of a pass-through, the subsidy level was expected to increase 46 percent in the MIS, which triggered AER's decision to increase electricity tariffs for commercial and industrial users at the end of 2016. Taking into account the likely international prices of liquefied natural gas (LNG) for Oman's market (the Asia-Pacific) and traded diesel prices, a recent International Renewable Energy Agency report calculated that the total volume of subsidies for power consumption in 2012 would amount to some $2.63 billion compared to the country's LNG export receipts of about $4 billion. The electricity sector has been restructured and regulatory reforms have been successfully implemented along with the transparent calculation of subsidies by AER. Much needs to be done in restructuring retail electricity tariffs and adjusting them upward. In 2014, the government subsidy was at 38 percent of the economic cost. The subsidies in the much smaller systems of DPC and RAECO in 2014 were 44 percent and 78 percent, respectively, of their economic costs. The government is working to gradually adjust the electricity tariffs to phase out the subsidy, but the impact mitigation mechanism is yet to be determined. More recently, effective January 2017, electricity tariffs were also increased for commercial and industrial customers.

Oman is currently developing and implementing a competitive wholesale market for electricity that will provide a route to market for generators and the creation of an electricity spot market. Generators not under a PPA contract will need to adapt to operating in such an environment.

Demand-side response is expected to play an increasing role in the mid to long term in Oman. By changing the profile of demand, and increasing its flexibility, the demand-side response can reduce the need for investment in generation and network capacity. Within this context, the new tariff will reflect the actual cost of supplying electricity and would provide a relatively small number of customers with strong incentives to reduce demand at peak times. This promises significant benefits in terms of reducing overall peak demand and the requirement for future investments.

Finally, stand-alone reverse osmosis will play a significant part in water production in the future. This has the potential to reduce the linkage between water and power production.

Conclusion

This chapter analyzed the performance of Omani electricity utilities in 2013. Oman's unbundling of the electricity sector has set the country on a path to overall reform in the power sector. The main challenges observed across utilities in Oman are related to covering their operating costs, which makes them heavily dependent upon state subsidies. Because revenues from the sale of electricity do not cover the total economic cost of supply, the Ministry of Finance provides an electricity subsidy to licensed suppliers on an annual basis. On the demand side, growing income per capita, continued government investment in infrastructure projects, and a growing population are all expected to contribute to a continued high growth in electricity demand in the sultanate.

Of the 12 GUs studied in this chapter, 8 are private (1 big, 4 small, and 3 medium) and 4 are public (2 small and 2 medium). With a fuel mix exclusively based upon natural gas, the GUs in Oman use some of the most efficient technologies in the region for fossil-fuel-based electricity generation. Fuel costs' share of total OPEX is close to the MENA median of 75 percent. The number of employees in GUs remains very low in comparison to other MENA economies, which results in high OPEX per employee (from $289,000 to $942,000). All utilities recover their OPEX from the sales of energy, which could be a result of OPWPC purchasing all the electricity from the GUs based upon PPAs. OPEX recovery values range from one to three times the MENA median. This is also reflected in the profitability indicators, which show positive performance for ROA and ROE across all the GUs for which data were available.

All three DUs in Oman are public: two medium and one small (MEDC). In terms of technical performance, there is still room for improvement, particularly for MJEC and MZEC, which have distribution losses higher than the MENA median value of 10 percent. The three utilities seem to perform well financially, with ROA and ROE values more than twice the MENA median in most cases. This seems to contradict the low performance reported on indicators such as accounts receivable, collection rate, and recovery of OPEX from sales, which would be expected to indicate poor overall financial performance. This could be explained by the fact that DUs in Oman benefit from transfers from other sources, such as the government, to help maintain their positive financial performance.

Unbundling and private sector involvement have contributed to the overall improvement of utility performance on several levels, though more so in generation activities than in distribution-related activities. Any analysis of Oman's power sector must consider the role that many electricity utilities play in desalination and water-related activities. On the supply side, the country still depends exclusively on gas and to a much lesser degree on diesel as sources of energy (about 2 percent of the energy mix), whereas the development of renewable energy—in spite of the abundant solar potential—has yet to be considered. However, the presence of subsidies, as in most parts of the MENA region, does not make renewable energy an attractive economic alternative for the generation of electricity, for which a review of tariffs and subsidies would need to take place. Not only would this encourage renewable energy development, but also it would allow for the country's energy sector to cope with increasing demand while minimizing the risks and impacts of higher fiscal burdens.

Last but not least come the issues of data collection and data quality. Across power sectors in MENA, Oman's may have the most transparent reporting, and the AER collaborated closely with us on this study. In terms of data quality, all DUs and the TU report having implemented supervisory control and data acquisition (SCADA), and a majority of utilities have adopted international accounting standards (IAS). As of 2013, more than half the GUs, all the DUs, and the two VIUs had yet to implement cost-accounting systems.

Notes

1. In addition, the Public Authority for Electricity and Water (PAEW) is the regulator for the water sector in Oman. Created in 2007, its role in the electricity sector is limited to policy overview. In addition to this, PAEW is also a direct water service provider.
2. Calculated from data on individual capacity of different plants (AER 2014, 9).
3. The indicators energy sales/total OPEX, energy sales/total costs, and accounts receivable are not applicable to generation utilities based on how these indicators were categorized for the purpose of the MENA Electricity Database. However, for comparative purposes, their values are presented and discussed in this chapter but not in previous chapters of this book.

References

AER (Authority for Electricity Regulation). 2014. *Annual Report 2014*. http://www .aer-oman.org/pdfs/Annual%20Report%202014%20-%20Eng.pdf.

AUE (Arab Union of Electricity). 2013. *Annual Statistical Bulletin 2013*. Amman, Jordan: AUE.

OPWPC (Oman Power and Water Procurement Company). 2013. *Annual Report 2013*. http://www.omanpwp.com/PDF/OPWP%20Annual%20Report%202013%20eng.pdf.

———. 2014. *Annual Report 2014*. http://www.omanpwp.com/PDF/01-AR-2014 -OPWP-Eng%20(12).pdf.

CHAPTER 10

Synopses of the Case Studies

Introduction

The four case studies presented (of the Arab Republic of Egypt, Jordan, Morocco, and Oman) offer insights relevant to the Middle East and North Africa (MENA) region and beyond. The studies aimed at providing not only an overview of each country's power sector but also an analysis of utility performance to help identify potential areas of improvement. This chapter presents the key findings of each case study.

Arab Republic of Egypt: An Urgent Need for Sector Reforms

The Egypt case study should be put in the context of 2013—the year of the data used for this analysis—recognizing that the Egyptian power sector has gone through some important changes since then.

The technical and operational performance of Egypt's generation utilities (GUs) is consistent with the Middle East and North Africa (MENA) median, but the country's commercial and financial performance is much worse. Capacity factors range from 58 percent to 70 percent of their full capacity, and availability factors range from 79 percent to 91 percent, which is consistent or better than the MENA median. But operational expenses (OPEX) per employee are lower, and costs are high because of high fuel costs and excessive staffing. Cost-recovery and accounts receivable indicators point to a major dependence on subsidies. Accounts receivable of Egyptian GUs are from 6 to 15 times higher than the MENA median, and none of the GUs in Egypt recover their total OPEX or their total costs from sales, except the Middle Delta Electricity Production Company, which recovers its OPEX, yet not its total costs.

This poor commercial performance explains the high fiscal cost of the sector: $1.6 billion in subsidies in 2013. For most of the GUs, the low cost- recovery reflects low selling tariffs, combined with the high cost of fuel, as well as in some cases low production and hence low sales. The need to borrow to finance business explains the very high debt-to-equity ratios of Egyptian GUs, between 4 and 10 times higher than the MENA median of 357 percent. It also explains the low current ratio, which is below 70 percent for all GUs. The outcome of the poor commercial and financial management of the GUs is a return on assets (ROA) and a return on equity (ROE) close to 0 percent in most cases, well below the 3 percent and 7 percent median value for ROA and ROE in MENA.

Meanwhile, Egypt's DUs have significant room for improvement on the technical and financial front but do reasonably well on the commercial dimensions of performance—in spite of the complex political and social context in which they need to operate. They also perform relatively well technically compared to their peers, although not on all dimensions. Their load factor and distribution losses are close to the regional MENA median. Their high nontechnical losses (owing to theft and erroneous meter readings) are, however, quite high and explain about 25 percent of total distribution losses. OPEX per employee is much lower than the MENA median, reflecting the largest number of employees in the region. The share of labor costs in total OPEX is two to three times higher than the MENA and non-MENA medians.

The high costs and the social context help explain why average tariffs were below costs in 2013 generally, further fueling the subsidy cost of the sector already noted for GUs. Some of Egypt's DUs get additional revenue. For instance, the Canal Electricity Distribution Company gets subsidies for the electricity exported to Gaza. Egypt's DUs compensate for the low cost-recovery rates and some of the excess costs resulting from their technical performance with a solid commercial performance in a difficult social context. They manage to enjoy high billing levels and high collection rates, close to or above the regional median. The only significant issue is the high receivable period (for example, almost six months for the North Cairo Electricity Distribution Company). This can be attributed to delayed collection cycles resulting from time-consuming manual registration of readings and billing. Since 2013, the Egyptian Electricity Holding Company has been exploring the option of shifting to smart meters as a potential solution to reducing nontechnical losses and the time taken for bill collection. Low cost-recovery is compensated not only by subsidies but also by borrowing. This explains the high debt-to-equity ratio and the low ratio of current assets to current liabilities. And this also explains the low ROE and ROA, except for Canal Electricity Distribution Company and South Cairo Electricity Distribution Company, which benefit from additional revenue sources.

Delaying important investment and management decisions because a country is going through difficult political and social times may simply lead to more political and social tension. While Egypt's reforms have allowed its utilities to achieve a technical and operational performance largely consistent with or often better than MENA's median performance, they have not been able to attract the investment needed to meet growing demand. Moreover, the sector has relied extensively on subsidies to finance its operational expenditures, a practice that is unlikely to be sustainable as the country adjusts its fiscal balance. Delaying investment decisions further could exacerbate political tensions by increasing the risk of consumer rationing.

Simply adding more installed capacity is not the only way to address growing demand. The country has been slow to cut costs and improve its commercial performance. Improving labor efficiency and cost-recovery should be on the agenda eventually, even if doing both jointly may be too difficult amid current tensions. Additional options include a redesign of tariff regulations to increase the scope for cross-subsidies and improve the social targeting of electricity pricing.

Jordan: Harvesting Results from a Restructuring of the Power Sector

The Jordan case study should be put in the context of 2013—the year of the data used for this analysis—recognizing that the Jordanian power sector has gone through some important changes since then.

Overall, Jordan's GUs do quite well on most performance dimensions for which data are available, with cost-recovery rates a notable outlier. They do not do very well on monitoring and transparency, because there are significant data gaps for some of the utilities. But the data are solid enough to be able to provide a strong diagnostic. At the technical and operational levels, the kingdom's GUs stand out by the high dispersion of their performance. The capacity factor is well below the MENA median for four of them, including two independent power producers (IPPs) that were not fully in operation in 2013, and well above for two of them (both are private—Amman East Power Plant and Qatrana Electric Power Company—with strong contractual service obligations). The availability factor is only obtainable for half the utilities but is consistent or better than the MENA median. At the operational level, Jordan's DUs are overstaffed, although the IPPs less so. Because Jordan is one of the few countries in the region that does not provide fuel subsidies, OPEX values[1] are much higher than values observed elsewhere in the region. This results in high shares of fuel costs for the GUs, ranging from 95 percent to 98 percent. Consequently, the share of labor costs in OPEX is considerably smaller than the MENA median, and OPEX per employee is much larger than the MENA median.

At the commercial level, only one utility recovered its total OPEX or total costs. The differences for the rest are essentially covered by subsidies. At the financial level, Jordan's GUs align with MENA values. The debt-to-equity ratio is similar across GUs and in some cases lower than the MENA median, with the exception of the fully state-owned utility Samra Electric Power Generating Company (SEPCO) at 876 percent. SEPCO has been expanding its generation capacity since 2010 by adding new generating units, of which most have been financed mainly through debt rather than equity. The GUs are all profitable, with high ROE and ROA values—in particular the IPPs—reaching at least three times the median MENA ROE of 5 percent. This is mainly a result of the provisions of the IPP contracts under which they operate, which ensure their profits.

Jordan's DUs are profitable. Compared with their MENA peers, they perform better on billing and cost-recovery but underperform on most other dimensions.

At the technical level, Jordan's three DUs have lower load factors and somewhat higher losses than the median MENA values. At the operational level, OPEX per employee figures for Jordanian utilities are much higher than the MENA and non-MENA medians. This is most likely due to the high OPEX figures attributed to other costs, because share of labor costs represents only about 6 percent or 7 percent of OPEX in each of the utilities. Differences in operational performance are influenced by the differences in the geographical areas covered by the utilities.

Total billing per connection is much higher than the MENA median. Differences across DUs in Jordan are explained by differences in the customer base. For instance, Jordan Electric Power Company benefits from higher consumption in the capital, Amman, and among its industrial consumers. The other utilities have more rural clients. Despite their good billing performance, only the Irbid District Electricity Company (IDECO) recovers its OPEX from sales. As in other countries of the region, all Jordanian DUs have high debt-to-equity ratios, suggesting a high level of financing through debt. Considering Jordan's DUs' underperformance on various indicators, it may be surprising that the bottom line is so positive and that Jordan's DUs enjoy high ROEs and ROAs relative to the regional medians. But this is partially linked to the fact that the licenses granted to the two privatized utilities, Electricity Distribution Company and IDECO, guarantee a 10 percent profit on their regulatory asset base after the regulator reviews and approves their annual budgets, their projects, and the anticipated electricity losses.

The power sector's main difficulty in 2013 may have been its excessive reliance on explicit and implicit subsidies as indicated by its high quasi-fiscal deficit. Because much of this is related to underpricing, it is important to get the pricing signals right. This helps managing demand. The main difficulty is that this would have to be done in a politically sensitive context: the poor, including many refugees, will have to be protected as much as possible from brutal price shocks while efforts to cut the sector's fiscal costs are implemented. Average tariffs can help ensure financial and fiscal viability. But tariff composition matters just as much to equity as the social, financial, and political viability of energy prices. Finally, to do the job right, a regulator critically needs information.

Morocco: Benefits and Challenges of Multiservice Providers

The case study of Morocco should be put in the context of 2013—the year of the data used for this analysis—recognizing that the Moroccan power sector has gone through some important changes since then.

From a technical perspective, Morocco's biggest private GU (Jorf Lasfar Energy Company [JLEC]) does much better than its MENA peers, with high capacity and availability factors. This is probably because in its power purchase agreement (PPA), electricity purchase is guaranteed and the utility is encouraged to maximize utilization of its generation capacity. The large multi-utility company, the Office National de l'Electricité et de l'Eau Potable (ONEE), is below par, with a capacity factor of 31 percent, suggesting that ONEE's facilities are operated as load followers and for peaking. The availability factor of ONEE's generation facilities, at 53 percent, can mainly be explained by temporary repair and maintenance issues. At the operational level, JLEC's cost performance is strong and the share of energy purchases and cost of fuel, lubricant, gas, and coal in total OPEX are high (94.5 percent).

The sector's commercial performance is excellent by any standard as indicated by cost-recovery and receivables. The cost-recovery rate in the case of JLEC is comfortably high (153 percent for recovery of OPEX from sales), as would be expected from an IPP. This level reflects the strong profitability of the business. Morocco's private power producers enjoy attractive contractual arrangements, and PPAs are designed to pass most market and institutional risks to ONEE. The cost-recovery performance of ONEE's generating activity is difficult to assess without detailed analysis of the overall costs of the vertically integrated utility (VIU), its sales, and the extensive use of cross-subsidies. The only hard evidence is that ONEE's electricity sales alone were not sufficient in 2013 to fully recover total costs. At 45 days, JLEC's receivables are almost the same as MENA's 40 days median. However, ONEE is quite high at 159 days.

The financial management is quite reasonable by MENA standards for the private GUs but unsustainable for ONEE. The debt-to-equity ratio of JLEC is 277 percent, which is relatively high but smaller than for many GUs in the region. It is much worse for ONEE (1,240 percent). At 17 percent, JLEC's ROE was high, while its ROA was 4 percent. ONEE was not profitable in 2013, with an ROA valued at −4.40 percent. Morocco is working on addressing the issue, but it is not an easy challenge because a large proportion of risks are passed to ONEE (for example, fuel price and exchange rate) to protect private GUs.

An assessment of DUs in Morocco is particularly challenging because they tend to deliver both electricity, water and sanitation services. In addition, all the small DUs are public (municipal distributors) except for AMENDIS Tetouan, which is private, whereas three of the five medium utilities are private. Accounting for this limitation, several characteristics can be sketched out. Technically, the DUs' load factors are all close to the MENA median value of 60 percent. Distribution losses in Morocco are lower than or equal to the MENA median of 10 percent.

Operationally, the DUs are not doing as well as their peers. While labor costs of all DUs make up 8 percent to 14 percent of OPEX, the OPEX per employee values are generally higher than the MENA median. This assessment is, however, biased by the fact that key information is not available for the private operators. The share of labor costs in total OPEX remains low for the utilities for which values were reported, ranging from 8 percent to 14 percent. This suggests that the OPEX is mainly made up of other costs, such as the costs of purchasing electricity from ONEE. OPEX per connection, per kilowatt-hour (kWh), and per kilometer (km) are almost all higher than the MENA medians. The measures are, however, imperfect because these companies' coverage of both electricity and water blurs key information on cost allocation. Moreover, electricity services are often used to cross-subsidize the water services and the heavy investments required in sanitation-related activities. Hence, tariffs and prices are often unrelated to sector-specific costs. They can include fees used by the operators to compensate for low national tariffs.

From a commercial perspective, Morocco's utilities do reasonably well with OPEX because for most of the DUs they are fully or almost recovered. Full cost-recovery is not the norm, but the country is working on closing the gap. Morocco has indeed worked on tariff adjustments designed to improve cost-recovery, while not affecting households with monthly electricity consumptions below 100 kWh. This offers a model for many other countries in the region.

From a financial perspective, all private utilities show strong profitability ratios with the exception of AMENDIS Tetouan, which shows negative ROA and ROE. This can be explained by the fact that AMENDIS Tangier and Tetouan are a joint concession, and the private operator compensates Tetouan with the business in Tangier. Data availability was the main obstacle to analyzing the financial performance of the municipal distributors.

Morocco's experience adds to the evidence provided by the other countries on a limited availability of access to performance indicators. Very little information on Morocco's electricity utilities is publicly available. The issue of data quality is also of concern, particularly for public DUs.

Oman: A Remarkably Sophisticated Power Market

The case study of Oman should be put in the context of 2013—the year of the data used for this analysis—recognizing that the Omani power sector has gone through some important changes since then.

At the technical and operational levels, performance varies across Oman's GUs. The capacity factors vary between 32 percent and 69 percent in the sample. Availability factors are notably high, ranging from 85 percent to 96 percent. This is because all the generating plants on the main interconnected system use gas turbines to generate electricity. Some of the differences in performance can be explained by the fact that a portion of the generation plants are also involved in desalination activities and that a few are used only for peaking. Most are private entities, which may explain why they have relatively small staff numbers and higher OPEX per employee than the region's median. The cost of gas accounts for a significant share of OPEX, representing as much as 78 percent. In general, fuel costs' share of total OPEX is similar or lower than the MENA median of 75 percent. Oman's GUs rely on a fuel mix exclusively based on natural gas and use some of the most efficient technologies in the region for fossil-fuel-based electricity generation.

At the commercial level, almost all utilities recover their total OPEX from the sales of energy. This good performance is, partially at least, the result of the Oman Power and Water Procurement Company purchasing all the electricity from the GUs based upon PPAs. OPEX recovery values range from one to three times the

MENA median. The strength of the commercial performance is confirmed by the relatively short average time lapse between accounts receivable and sales, guaranteeing a constant flow of cash on hand.

As in many countries of the region, financial performance is an issue. The debt-to-equity ratio of the GUs in Oman varies from 72 percent to 94 percent to 357 percent. For utilities with a small ratio, the issue may be excess cash. For utilities with a high ratio, the explanation is an excessive tendency to borrow to finance projects and operations, even when cash is available. For most of the utilities, the current assets to current liabilities ratio is divided into those utilities with a ratio below 100 percent and those above. Al Batinah Power Company the Al Suwadi Power Company, Al-Kamil Power Plant, Barka Power and Desalination Plant, and the United Power Company showed values in the range of 38 percent to 54 percent. For these utilities, current liabilities are very high compared to the current assets and are, therefore, not liquid. The most puzzling indicators concern profitability. In view of the sector's strong commercial performance, it is surprising to see such low ROAs, even if they are quite in line with the MENA median. They range from 0.13 percent to 9 percent. The range for ROE is broader: 0.2 percent to 24 percent. The main drivers of this dispersion are availability and reliability. Changes in the demand and supply landscape will not affect profits, because there is a pass-through cost for the generators.

The three DUs in Oman are public. Our detailed analysis indicates average technical and operational performance, poor commercial performance, and very good financial performance by MENA standards—despite poor cost-recovery rates.

At the technical level, the load factors are low by MENA standards and losses are at the MENA median or worse. The diversity largely reflects differences in the customer base. At the operational level, OPEX per employee is average by MENA standards, although labor costs' share of OPEX is notably low, at half or less the MENA median. The OPEX per connection is above the MENA median, although there is no clear pattern for OPEX per km. Regarding commercial performance, the data available suggest average performance by MENA standards, and low by global ones. Accounts are received within 110–122 days, and collection rates are significantly lower than the region's median, with values ranging from 74 percent to 79 percent. The recovery of OPEX (and hence total costs) ranges from 61 percent to 80 percent, which highlights the importance of other forms of sector financing. In this case, however, debt plays a smaller role than in other countries of the region. With debt-to-equity ratios ranging from 109 percent to 148 percent, Oman's utilities seem to be doing much better than the MENA median. But this is not confirmed by the ratio of current assets to current liabilities, which ranges from 18 percent to 46 percent. This is low by any standard and hints at a limited ability to repay current liabilities from cash or assets. For now, this does not affect the profitability of the DUs, which benefit from public transfers to help maintain their positive financial performance. Muscat Electricity Distribution Company and Majan

Electricity Company have similar ROAs (8 percent each). Muscat also has the highest ROE (16 percent), followed by both Majan and Mazoon Electricity Distribution Company at 14 percent each. This seems to contradict the low performance reported by indicators such as accounts receivable, collection rate, and recovery of OPEX from sales, which should be expected to indicate poor overall financial performance.

Ultimately, the review of Oman's experience provides evidence that the significant restructuring adopted by the country is technically and institutionally feasible. In 2013, the main doubt was related to the financial sustainability of the model. If the subsidy policy was to be maintained, it seems likely that the expected growth in demand would increase fiscal pressure. Unless this risk is addressed through closer monitoring of financial and commercial performance, the relative and absolute costs of the sector will continue to grow, despite a strong reliance on private operators to finance specific needs. More data are essential to all these efforts.

Note

1. The cost of fuel was estimated for the Jordanian generation utilities based upon the average cost of fuel per kWh from the regulator and the kWhs generated by each utility in 2013. This cost of fuel was then added to the operating costs to obtain the total OPEX as per the definition used in this study.

Conclusion

The potential management and financing payoffs of the new MENA Electricity Database produced in the context of this analysis are hard to ignore. Because the database covers most power utilities in the Middle East and North Africa (MENA), it is broad enough to produce robust insights into the sector's achievements and its challenges.

Those insights, synthesized in this concluding section, can be turned into concrete management and policy decisions. But it should be remembered that the baseline year of the study is 2013, and the power sector has changed since then, in some economies more than others. An appropriate response, of course, is to expand and extend the analysis and data collection begun here. And this is what we hope each economy will decide to do working with its utilities. A fundamental lesson of this study is that data analysis is essential to performance diagnostics at the utility level and at the sector level.

Cutting Hidden Costs in the Power Sector Is Key to Financing Sorely Needed Investment

Explicit and implicit subsidies of MENA's power sector impose a very heavy burden on taxpayers and power users. The burden can be measured in the utilities' hidden costs, or quasi-fiscal deficits (QFDs), which express the cost of not operating in the manner of a well-run utility. The QFD encompasses four types of inefficiencies: collection losses, transmission and distribution losses, underpricing, and overstaffing.

Estimates of the power sector's QFD range between −0.1 percent of gross domestic product (GDP) for the West Bank to 8.9 percent in Lebanon. To put this in context, consider that in Sub-Saharan Africa, where social concerns are at least as large as in the MENA region, the sector's QFD ranges from −0.3 percent to 6.0 percent. Half of the 14 MENA economies studied have a QFD in excess of 4 percent of the entire economy's GDP. The QFD share of GDP is relatively small in Maghreb economies and large in some Mashreq and Gulf Cooperation Council economies. The median value of about 4 percent of GDP represents one and a half times the average investment needed in the region's electricity sector, estimated at about 3 percent of GDP. In other words, the sector's investment gap could be filled simply by halving the current level of inefficiency.

At the utility level, performance varies widely. When measured as a share of utilities' revenue, QFDs range from 25 percent for a West Bank distribution utility (DU), Northern Electric Distribution Company (NEDCO), to almost 1,300 percent for the vertically integrated Iraqi power ministry. The QFD of at least 13 utilities exceeds their revenue. These figures reveal the extent to which utility-specific inefficiencies common in the region may be preventing self-financing.

Underpricing Is the Major Source of Inefficiencies, Although Otherwise Inefficiencies Are Economy and Utility Specific

The inefficiencies reflected in the QFD are linked both to policy and management decisions. The sources of inefficiencies, and hence the nature of the solutions, vary across economies. About two-thirds of the QFDs we detected can be traced to tariffs being set below cost-recovery levels in most economies, which nearly always reflects a political decision intended to protect current users. Even under such circumstances, however, managing costs can go far to enhance revenues. For example, Jordan's high levels of cost inefficiency are due largely to electricity production costs that reflect the preponderant role of diesel and fuel oil in generation.

The remaining one-third is explained by commercial losses, collection failures, and overstaffing, which are all mostly management decisions, although overstaffing may sometimes represent a political decision if it is an issue for all utilities in a given economy. These sources of inefficiencies should not be underestimated, as they represent half of the resources needed for the sector's investment needs. Overstaffing is of particular concern in only a few utilities, almost all of them DUs in Egypt. Collecting bills seems to be a significant challenge for DUs in Djibouti, Jordan, and the West Bank. Technical losses are significant for two of the West Bank operators (Jerusalem District Electricity Company and NEDCO) and for the Republic of Yemen's vertically integrated utility (VIU).

Low tariffs and overstaffing often reflect good intentions, but they are not the most effective ways to ensure that the poor can afford electricity or to boost employment. Moreover, given their present macroeconomic prospects, many MENA economies cannot afford to continue to lavish on average 2 percent of GDP on poorly targeted electricity subsidies. Improving the sector's performance will allow economies to increase the social returns on fiscal resources by allocating savings where they will do the most good, whether within the sector or outside it.

Identifying and unbundling hidden cost drivers and inefficiencies at the utility level can pinpoint areas for improvement—whether financial, technical, commercial, or labor—and, from a regulatory perspective, improve the accountability of key actors. From the perspective of sector policy, quantifying the QFD provides governments with a rough order of magnitude of the improvements.

Taking advantage of readily available opportunities to reduce cost inefficiencies in the generation and distribution of electricity will also make the sector more sustainable and increase the creditworthiness of utilities, thus facilitating access to commercial financing.

MENA's Power Sector Must Match Its Technical Success with Improvements in Commercial and Financial Management

For more than half of the indicators selected—most of them technical—the region's economies tend to perform better than the sample of economies outside MENA. Unfortunately, there does not seem to be a clear correlation between good technical performance and sustainable financial performance, and unless the sector can increase its revenue or better manage its costs, the current technical level is unlikely to be sustainable (table CL.1).

On the technical and operational side, the international comparison and the trend analysis point to a significant increase in operating expenses (OPEX)

Table CL.1 Comparing Median Utility Performance in the MENA Region and Elsewhere

	All utilities	Distribution utilities	Vertically integrated utilities
Technical and operational			
OPEX/connection ($)	—	MENA higher	MENA higher
OPEX/kWh sold ($)	—	MENA lower	Samples too small
Residential connections/employee	—	MENA lower	MENA lower
Distribution losses	Equivalent	—	—
Commercial			
Energy sold (kWh)/connection	MENA higher	—	—
Total billing/connection	MENA somewhat higher	—	—
Collection rate	MENA somewhat lower	—	—
Financial			
Sales/OPEX (%)	—	MENA somewhat lower	MENA somewhat higher
Sales/total costs (%)	—	MENA higher (depending on subsidies)	MENA lower (depending on subsidies)
Accounts receivable/sales (days)	MENA much higher	—	—
Debt/equity	MENA much higher and essentially unsustainable	—	—
Current assets/current liabilities	Equivalent but not ideal	—	—
Return on assets (%)	MENA somewhat higher but not high enough to stimulate financing	—	—
Return on equity (%)	MENA higher but not commensurate with risk	—	—

Source: World Bank calculations.
Note: Comparisons are only made for all utilities together when the indicator has the same meaning for different types of utilities. Otherwise, comparisons are made separately for distribution utilities and vertically integrated utilities. kWh = kilowatt-hours; MENA = Middle East and North Africa; OPEX = operating expenses; — = not applicable.

Shedding Light on Electricity Utilities in the Middle East and North Africa
http://dx.doi.org/10.1596/978-1-4648-1182-1

during the period covered, which is consistent with the increase in oil prices from 2009 to 2013. On commercial management, the indicators reveal (a) a high dependence on subsidies to recover costs and (b) a high tolerance for nonpayment (with a ratio of accounts receivable to sales that is almost three times that of non-MENA economies). On financial dimensions, despite return-on-assets and return-on-equity values that are somewhat better than those of non-MENA peers, the region appears to be relying on a risky strategy as indicated by (a) a low ratio of current assets to current liabilities (lower than 100 percent) and (b) an exceptionally high debt-to-equity ratio (almost four times the non-MENA median), leaving utilities highly exposed to external shocks.

The importance of labor costs highlighted by the QFD analysis is likely to be a particularly sensitive topic in any policy discussion of the data reported here. In a region where underemployment is a major problem, it is impossible not to recognize the political sensitivity of efforts to improve labor indicators. Where the matter is so sensitive that overstaffing in the power sector simply cannot be broached, it may nevertheless be useful to quantify the costs of not addressing the issue, thus clarifying the implications for subsidy levels (if revenues cannot be increased).

Because partial indicators of utility performance can lead to heterogeneous rankings of utilities, we applied an "average rank score" methodology to help utilities assess their performance against other utilities across a set of relevant indicators. The average rank score makes it possible to identify the better-performing utilities within a group that share a common set of data and for which reliance on a single indicator could be misleading. The main takeaways from this diagnostic across utility types are as follows: (a) for generation utilities (GUs), the best-performing utility is Qatrana Electric Power Company (Jordan), followed by Al-Kamil Power Plant (Oman) and ACWA Power Barka (Oman); (b) for DUs, Electricity Distribution Company (Jordan) is the best-performing utility in the group, followed by LYDEC (Morocco) and Jordan Electric Power Company (Jordan); and (c) for VIUs, the best performance is by Saudi Electricity Company (Saudi Arabia), followed by Société Nationale de l'Électricité et du Gaz (Algeria).

Well-Targeted Institutional and Economic Reforms Would Boost MENA's Power Sector

The variety of organizational structures found in the electricity sector around the world is quite striking. This study reveals that the MENA region is no exception. Utilities are central to all the organizational models encountered in the region, but otherwise these models show substantial institutional and contextual differences, some of which have been credited with, or blamed for, differences in utilities' performance.

Our assessment of the correlations between various institutional and contextual characteristics (utility type, size, ownership, the presence of a separate regulatory agency, and national income) and performance indicators, despite limitations (notably the use of cross-sectional rather than time-series data), suggests how and where reform policies may be most effective. Of the 36 performance indicators used for this analysis, 25 showed impact for one of the characteristics; in 14 cases, more than one characteristic (or "driver") was statistically significant. The results support the hypothesis that performance differences between utilities are likely to be correlated with institutional and economic policy variables, although a more thorough analysis is needed to be able to establish causality.

Utility type and size are the policy-related drivers that were most often significant (each for 30 percent of the indicators tested), while ownership type (public or private) and the presence of an independent regulator were significant for about 20 percent of the indicators tested. National income level was significant in 35 percent of the tests, indicating that this variable should be considered in any comparison across economies.

The impacts of reform would not be felt across all indicators but are likely to be concentrated in certain aspects of performance. Table CL.2 shows that the significant results for each driver are concentrated within two or three categories of indicators. For example, utility type has a substantial proportion of significant links to the indicator categories of losses efficiency, profitability, and consumption and billing, and no links at all to the categories of systems and operational efficiency, cost structure, cost-recovery, balance sheet, and metering. Ownership and regulation are linked to cost efficiency and labor efficiency. This suggests that improvements in cost efficiency and labor efficiency are particularly susceptible to reform efforts, because ownership and regulation are relatively easy factors to adjust. Other categories of indicators may be influenced by other drivers or by a complex combination of factors that simple testing of one characteristic at a specific point in time was unable to duplicate.

The Case Studies Yield Valuable Insights on the Variety and Nature of Reform Paths

Egypt, Jordan, Morocco, and Oman are analyzed in detail in the case studies presented in part II. The four countries represent the diverse challenges faced by economies in the region, as well as different paths taken toward electricity reform over the past 10–15 years. The four countries are characterized by quite different economic and political environments, which affect the degree of ease or difficulty involved in implementing reforms.

Egypt has not enjoyed the political stability often needed when undertaking significant reforms. Its experience indicates that demand shocks linked to political tensions may have a much stronger impact on the sector's commercial and financial performance than on its technical and operational performance.

Jordan has had to address both a demand and a supply shock. On the supply side, it has been affected by the need to drastically change its sources of energy owing to a break in gas supply from its main supplier in 2012. On the demand side, it has had to deal with unexpected increases resulting from a large inflow of refugees. The case study illustrates the impact of efforts to significantly scale up the role of the private sector in absorbing these shocks.

Table CL.2 Tests of Equality between Subgroups of Factors Related to Indicator Mean Values (Probabilities) Using One-at-a-Time Testing, MENA Utilities

Classes of utilities included	Category	Indicator	Number	Mean	Utility type	Size	Income	Ownership	Separate regulatory agency present
VIU vs. DU	System and operational efficiency	Load factor	23	0.56	0.80	0.25	0.96	0.07*	0.63
VIU vs. GU		Capacity factor	20	0.54	0.07*	0.61	0.12	0.43	S
VIU vs. GU		Availability factor	11	0.93	0.50	0.71	0.04**	0.50	0.50
VIU vs. TU vs. DU		Network maintenance	10	0.02	0.79	0.41	0.85	0.52	0.20
VIU vs. DU		Share of meters replaced (%)	9	0.02	0.41	0.40	0.70	0.29	0.82
VIU vs. TU	Losses efficiency	Transmission losses	3	0.03	0.86	S	S	S	S
VIU vs. DU		Distribution losses	37	0.13	0.001**	0.76	0.52	0.63	0.69
VIU vs. DU		Technical losses	18	0.075	0.0003**	0.37	0.32	0.22	0.14
VIU vs. DU		Nontechnical losses	18	0.049	0.0003**	0.88	0.32	0.13	0.48
VIU vs. GU vs. TU vs. DU	Cost efficiency	OPEX/employee	48	274,000	n.a.	0.0001**	0.003**	0.006**	0.39
VIU vs. DU		OPEX/connection	36	723.0	n.a.	0.16	0.0001**	0.99	0.80
VIU vs. DU		OPEX/kWh sold	36	0.11	n.a.	0.002**	0.51	0.41	0.0001**
VIU vs. TU vs. DU		OPEX/km	37	24,381.0	n.a.	0.006**	0.95	0.02**	0.001**
VIU vs. DU	Labor efficiency	Residential connections/employee	24	238	n.a.	0.09*	0.54	0.15	0.02**
VIU vs. DU		Energy sales/employee	31	170,000	n.a.	0.03**	0.48	0.0007**	0.005**
VIU vs. DU		Total revenues/employee	34	212,000	n.a.	0.1*	0.70	0.004**	0.001**
VIU vs. GU	Cost structure	Cost fuels/OPEX	22	0.65	0.12	0.16	0.47	0.97	0.02**
VIU vs GU		Energy purchases + fuels/OPEX	8	0.77	S	0.05**	0.23	S	0.70
VIU vs. GU vs. DU		Labor cost/OPEX	35	0.13	0.22	0.03**	0.02**	0.13	0.29
VIU vs. DU	Cost recovery	Energy sales/OPEX	32	0.95	0.42	0.49	0.07*	0.83	0.15
VIU vs. DU		Energy sales/costs	19	0.82	0.11	0.13	0.03**	0.54	0.48

table continues next page

Table CL.2 Tests of Equality between Subgroups of Factors Related to Indicator Mean Values (Probabilities) Using One-at-a-Time Testing, MENA Utilities *(continued)*

Classes of utilities included	Indicator	Category	Number	Mean	Utility type	Size	Income	Ownership	Separate regulatory agency present
VIU vs. DU	Accounts receivable	Balance sheet	26	161	0.11	0.22	0.06*	0.84	0.63
VIU vs. GU vs. TU vs. DU	Debt/equity		47	7.08	0.24	0.05**	0.04**	0.62	0.67
VIU vs. GU vs. TU vs. DU	Assets/liabilities		53	1.17	0.32	0.0005**	0.31	0.56	0.84
VIU vs. GU vs. TU vs. DU	Return on assets	Profitability	49	0.3%	0.39	0.07*	0.22	0.05*	0.40
VIU vs. GU vs. TU vs. DU	Return on equity		46	4.6%	0.009**	0.10	0.15	0.03**	0.12
VIU vs. DU	Total energy volume/connection	Consumption and billing	35	6.4	0.002**	0.36	0.001**	98.0	0.21
VIU vs. DU	Residential energy volume/connection		23	4.0	0.01**	0.72	0.0001**	0.62	0.51
VIU vs. DU	Total billing/connection		27	297	0.17	0.005**	0.0001**	0.037**	0.09*
VIU vs. DU	Residential billing/connection		22	258	0.59	0.0001**	0.007**	0.37	0.34
VIU vs. DU	Collection rate		15	88%	0.03**	0.003**	0.86	0.51	0.08*
VIU vs. DU	Share of installed meters (%)	Metering	15	96%	0.32	0.33	0.02**	0.72	0.75
VIU vs. TU vs. DU	SAIFI	Customer management and service quality	15	1.6	0.02**	0.70	0.06*	0.37	0.69
VIU vs. TU vs. DU	SAIDI		12	28.6	0.46	0.35	0.72	0.49	0.57
VIU vs. TU vs. DU	CAIDI		9	52	0.21	0.46	S	S	0.20
VIU vs. TU vs. DU	Duration of interruptions		5	2.0	S	0.99	0.03**	0.32	0.03**

Source: World Bank calculations.

Note: Significant results are shaded in light red; performance indicators for which more than one factor gave significant results in one-at-a time testing are shaded in blue. CAIDI = Customer Average Interruption Duration Index; DU = distribution utility; GU = generation utility; km = kilometer; kWh = kilowatt-hour; MENA = Middle East and North Africa; n.a. = not applicable (tests are inappropriate); OPEX = operating expenses; S = singular dataset so estimation is not possible; SAIDI = System Average Interruption Duration Index; SAIFI = System Average Interruption Frequency Index; TU = transmission utility; VIU = vertically integrated utility.

Significance level: * = 10 percent; ** = 5 percent.

Morocco illustrates how electricity reforms can be implemented in a hybrid market in which regional utilities cover electricity as well as water and sanitation. This peculiarity makes it difficult to differentiate the allocation of resources across the two activities but does allow for the introduction of cross-subsidies.

Finally, Oman is a relatively small economy where policy reforms have eased access to private financing in the sector. It now has long experience with an unbundled electricity sector. Private GUs are also involved in the desalination efforts that ensure the sultanate's water supply.

More Systematic Monitoring of Power Sector Performance Is Needed

The MENA Electricity Database can be used not only to produce a snapshot of the region's power sector but also to clarify the managerial, technical, and policy steps that might be required to meet fast-growing demand from all economic actors, including residential users. Just as important, and perhaps more subtly, the database provides a baseline against which future progress can be tracked and measured. To be effective and to ensure accountability of policy makers and managers, progress needs to be measured from baseline to target, which is how comparisons can become an input for policy. Targets are best set at the firm level for most operational matters, but sector-level targets are needed as well if governments are to address the fiscal and social concerns and constraints raised in the analysis.

The new database produced for this report offers the region access to a comparable dataset for a statistically significant sample of economies both within and outside the region. The comparable components of the dataset cover indicators in three broad performance categories: (a) technical and operational, (b) financial, and (c) commercial. But the dataset also exposes the monitoring weaknesses of the region. Very little comparable information exists for GUs, for example. On many performance indicators, comparability is not possible, either for lack of data or because the indicators have different meanings for the different types of utility.

The gaps in the data needed for good policy and management are real but not unsurmountable. To help fill them, authorities in the region may wish to impose on regulated industries guidelines and other information-sharing requirements derived from modern regulatory practice. For unregulated companies, standard accounting reports and annual balance sheets can go a long way toward supplying the raw data needed to improve monitoring of the region's electricity sector, provided the will to use that information is present.

Without a political commitment to improve the dataset and to use it to monitor progress and fine-tune policy, it will be difficult for the sector's decision makers to track efforts to cut the sector's financing deficits and close its service gaps. The analysis provided here has shown how much room there is to cut specific costs and to enhance revenue. It has also shown, for many economies in the region, the unsustainability of a business-as-usual approach. Without the checks and balances provided by an effective monitoring system, progress in addressing challenges cannot be tracked adequately. The case for change in the region's monitoring practices is thus strong—and change is possible. Many policy makers are already moving in the right direction by making important institutional changes. How fast and how intensively they move is likely to determine how quickly the financing and service needs of the sector are met.

Manual of Indicators and Data Sources

MENA Electricity Database

The MENA Electricity Database covers 67 utilities in 14 economies of the Middle East and North Africa (MENA) region. This information was gathered through a questionnaire sent to electricity utilities, in which they were asked if they performed one or several of the following activities: generation, transmission, and distribution. According to their responses, the utilities were categorized as vertically integrated utilities (VIUs), generation utilities, distribution utilities, or transmission utilities.

Detailed data collected through the questionnaire enabled the calculation of 36 indicators of the utilities' performance across two key measures of efficiency (operational and financial) and two measures of service quality (technical and commercial). Data were obtained for the years 2009–14, making it possible to follow the evolution of the performance of one specific utility.

Core Performance Indicators for 67 Utilities in the MENA Region

A list of *key data* that would be required to calculate a set of *priority indicators* was identified. An attempt was made to focus on primary data, that is, data that would be commonly generated through internal processes and reported to the utility's management or contained in standard financial, technical, and commercial reports.

The indicators selected are those that can provide insights into key technical, commercial, and financial elements of utility performance. These indicators are commonly used by the industry, regulators, and academic and international organizations to assess the different dimensions of utility performance, taking into account key specificities.

The final choice of data and indicators was based on a review of international experiences of similar data collection and benchmarking exercises. Various reports related to global and regional initiatives were reviewed, as well as similar initiatives and programs of national or local energy regulators such as the Office of Gas and Electricity Markets (OFGEM, United Kingdom), the Ontario Energy Board (OEB, Canada), and the Australian Energy Market Commission (AEMC), as well as specialized reports and analysis from international consultants (for example, Hesmondhalgh and others 2012). It was also ensured that major *inputs* and *outputs* required for statistical benchmarking analysis would be available if our information requests were satisfied.

Table A.1 contains a total of 36 indicators that are grouped into (a) 16 technical and operational indicators; (b) 10 financial indicators; and (c) 10 commercial indicators, which are further subdivided into four, four, and three subgroups respectively.

Table A.1 Descriptions of the 36 Core Indicators

Name	Unit	Description	Subactivity
Technical and operational indicators			
System and operational efficiency			
Load factor	%	(Energy delivered to distribution network in MWh/8,760 hours)/maximum demand on the interconnected system in MW	**DU-VIU**
Capacity factor	%	(Total net generation in MWh/8,760 hours)/total installed generation capacity in MW	**GU-VIU**
Availability factor	%	[(Total installed generation capacity * 8,760) − (total capacity hours out of service)] * 100/(total installed generation capacity * 8,760)	**GU-VIU**
Network maintenance	%	Length of existing network subject to major repair or replacement/(length of transmission network + length of distribution network)	**TU-DU-VIU**
Number of meters replaced/total number of meters	%	Number of meters of existing connections replaced/total number of meters	**DU-VIU**
Losses efficiency			
Transmission losses	%	Energy lost during transmission of power as a percentage of the sum of the total net energy generated and the energy purchased	**TU-VIU**
Distribution losses	%	Energy lost during distribution of power as a percentage of the sum of the total net energy generated and the energy purchased	**DU-VIU**
Technical losses	%	Distribution losses due to the technical characteristics of the distribution network	**DU-VIU**
Nontechnical losses	%	Distribution losses due to unmetered and unbilled consumption due to illegal connections, inaccurate estimations of consumptions, billing errors	**DU-VIU**

table continues next page

Table A.1 Descriptions of the 36 Core Indicators *(continued)*

Name	Unit	Description	Subactivity
Cost efficiency (total OPEX)			
Total OPEX/FTE employee	$/employee	Total OPEX/(number of FTE own employees + number of FTE employees from outsourced contracts)	**GU-TU-DU-VIU**
Total OPEX/connection	$/connection	Total OPEX/number of connections	**DU-VIU**
Total OPEX/kWh sold	$/kWh	Total OPEX/energy billed (excluding exports)	**DU-VIU**
Total OPEX/km of network	$/km	Total OPEX/(length of transmission network + length of distribution network)	**TU-DU-VIU**
Labor efficiency			
# of residential connections/FTE employee	Connections/ employee	(Total number of residential connections)/(number of FTE own employees + number of FTE employees from outsourced contracts)	**DU-VIU**
Energy sales ($)/FTE employee	$/employee	Total sales in $ related to energy service (consumption + fixed charges)/(number of FTE own employees + number of FTE employees from outsourced contracts)	**DU-VIU**
Total revenues ($)/FTE employee	$/employee	Total utility's revenues in $/(number of FTE own employees + number of FTE employees from outsourced contracts)	**DU-VIU**
Financial indicators			
Cost structure			
Share of cost of fuel, lubricant, gas, and coal in total OPEX	%	Cost of fuel, lubricant, gas, and coal/total OPEX	**GU-VIU**
Share of energy purchases and cost of fuel, lubricant, gas, and coal in total OPEX	%	(Cost of fuel, lubricant, gas, and coal + energy purchases)/ total OPEX	**VIU**
Share of labor cost in total OPEX	%	Labor cost/total OPEX	**GU-TU-DU-VIU**
Cost-recovery			
Energy sales/total OPEX	%	Revenues related to energy consumption and service in $/OPEX in $	**DU-VIU**
Energy sales/total costs	%	Revenues related to energy consumption and service in $/(OPEX + depreciation of fixed assets + other depreciation and provisions − net interests)	**DU-VIU**
Balance sheet			
(Accounts receivable/ sales) * 365	Days	(Accounts receivable at year end/energy sales in $) * 365	**DU-VIU**
Debt/equity	%	Total debt at year end in $/total equity	**GU-TU-DU-VIU**
Current assets/current liabilities	%	Total of current assets/total of current liabilities	**GU-TU-DU-VIU**
Profitability			
Return on assets	%	Net profit of the year/net fixed assets at year end	**GU-TU-DU-VIU**
Return on equity	%	Net profit of the year/total equity	**GU-TU-DU-VIU**

table continues next page

Shedding Light on Electricity Utilities in the Middle East and North Africa
http://dx.doi.org/10.1596/978-1-4648-1182-1

Table A.1 Descriptions of the 36 Core Indicators *(continued)*

Name	Unit	Description	Subactivity
Commercial indicators			
Average consumption and billing			
Total energy volume sold (kWh)/connection	kWh/ connection	Total sales of energy in kWh/# of connections	**DU-VIU**
Residential energy volume sold (kWh)/ connection	kWh/ connection	Residential sales in kWh/# of residential connections	**DU-VIU**
Total billing ($)/ connection	$/connection	Total sales related to energy service (consumption + fixed charges)/# of connections	**DU-VIU**
Residential billing ($)/ connection	$/connection	Residential energy sales in $/# of residential connections	**DU-VIU**
Collection rate	%	Income effectively collected during year/income billed	**DU-VIU**
Metering			
Share of installed meters	%	# of meters/# of connections	**DU-VIU**
Customer management and service quality			
SAIFI	Thousands	Takes into account interruptions affecting customers due to planned and unplanned events	**TU-DU-VIU**
SAIDI	Minutes	Takes into account interruptions affecting customers due to planned and unplanned events	**TU-DU-VIU**
CAIDI	Minutes	Tracks interruptions due to planned and unplanned events	**TU-DU-VIU**
Duration of interruption taken into consideration for system interruptions affecting customers (for example, SAIDI, SAIFI, and CAIDI customer measures)	Minutes	Corresponds to the minimum duration of interruptions in the consumer-side customers that are considered toward customer reliability indicators (planned and unplanned events)	**TU-DU-VIU**

Source: World Bank calculations.
Note: CAIDI = Customer Average Interruption Duration Index; DU = distribution utility; FTE = full-time equivalent; GU = generation utility; km = kilometer; kWh = kilowatt-hours; MW = megawatts; MWh = megawatt-hours; OPEX = operating expenses; SAIDI = System Average Interruption Duration Index; SAIFI = System Average Interruption Frequency Index; TU = transmission utility; VIU = vertically integrated utility.

Data Collection: Results and Challenges

Data Results and Sources

The main source of data for the MENA Electricity Database was an elaborate questionnaire developed for the purpose of the study. Supplementary data were obtained from the utilities' annual, financial, and activity reports, as needed.

For the non-MENA comparative portion of the database, data were obtained for 181 economies from several sources, mainly from the following:

- European countries from the AF-MERCADOS EMI report, 2013 (except for Denmark)
- African utilities from the Africa Infrastructure Country Diagnostic, 2005; (see Eberhard and others 2008)

- African utilities from Africa Power Subsidies, 2010; (see Trimble and others 2016)
- Latin American utilities from the Commission for Energy Regulation (CER) report, 2010; (see Andres and others 2012)
- Latin American utilities from the Latin America and Caribbean (LAC) Database, 2007; (see Andres and others 2012)
- Australia, Denmark, Sri Lanka, and Vietnam from Readiness for Investment in Sustainable Energy, 2013; (see Banerjee and others 2016)

Following is a summary of the number of relevant observations obtained for the non-MENA utilities, by type of utility (table A.2). A total of 1,041 indicator points were obtained for the 36 key indicators. The dominant type in our sample is distribution utilities, most of them from the LAC region.

Regarding the collection of data on MENA, two main points are to be highlighted:

- *Relevance of indicators to the type of utility*

Some indicators are relevant only to certain utility types. For example, distribution losses are experienced by both vertically integrated and distribution utilities but are not relevant to generation or transmission utilities. Other indicators, such as those relating to operating expenses (OPEX) are not comparable between structure types—the nature of the distribution business is such that the OPEX of a distribution utility would be much lower than that of a VIU, which includes some generation to serve the same number of customers.

- *Availability of indicators (and data used to calculate indicators) from the utilities*

Because not all these utilities actually collect the relevant data required in our data collection exercise, missing observations were commonly encountered. This means that for some indicators the number of available observations can be small.

Table A.2 Number of Indicator Points and Number of Utilities, by Type of Utility and Region

	Vertically integrated utility		Distribution utility		Generation utility		Transmission utility	
	Number of indicator points	Number of utilities	Number of indicator points	Number of utilities	Number of indicator points	Number of utilities	Number of indicator points	Number of utilities
Africa	124	25	27	5	0	0	0	0
Asia	7	1	40	9	4	2	2	2
Australia	0	0	13	1	0	0	0	0
Europe	0	0	12	3	8	2	5	2
LAC	46	9	721	117	0	0	0	0
Israel	5	1	0	0	0	0	0	0
United States	27	2	0	0	0	0	0	0

Source: World Bank calculations.
Note: LAC = Latin America and the Caribbean. "Africa" here refers to Sub-Saharan Africa.

Table A.3 Number of Indicator Points Collected, 2009–13

	2009	2010	2011	2012	2013
Total collected	703	744	817	873	1,164
Not applicable	93	97	111	124	187
Total collected of 1,600 applicable	610	647	706	749	977
Share of collected and applicable (%)	38	40	44	47	61

Source: World Bank calculations.

Table A.4 Number of Data Points Collected, 2009–13

	2009	2010	2011	2012	2013
Total collected	812	865	930	974	1,271
Not applicable	3	3	4	5	19
Total collected of 2,027 applicable	809	862	926	969	1,252
Share of collected and applicable (%)	40	43	46	48	63

Source: World Bank calculations.

Table A.3 presents the number of actual observations collected and available for the 36 core indicators used in this study. If all the 67 utilities were to ideally provide all the data required, we would have expected to obtain 1,600 indicator observations per year. Yet this is not the case.

The highest number of relevant (or applicable) observations were obtained for the year 2013, where 977 observations were obtained out of 1,600 (equivalent to 61 percent). These indicators were calculated from the data points collected (40 data points required for the calculation of the 36 core indicators), shown in table A.4.

Main Challenges

Most of the gaps encountered involved financial and technical data. The most common challenges encountered during the data collection process are presented below.

Different types of data. Different types of data were requested, that is, regulatory, financial, technical, and commercial. Therefore, responding to the questionnaire required the collaboration of different departments within the utility and the involvement of different senior executives who were not always accustomed to working together.

Format of financial information. The reporting format of utilities' financial statements differ depending on the legal environment (for example, if accounting plans and financial reporting formats are defined by local law or not) and accounting practices and philosophies (for example, the Anglo-Saxon accounting model versus the French or Spanish model). Many utilities have already or are progressively adopting international financial reporting standards (IFRS), but others have not done so yet.

Generally, the financial and accounting data requested here were to be easily identifiable in the main financial statements of the utility: balance sheet, income, cash flow, and use and sources of funds (the last two were not always available). When there was a doubt about a financial or accounting variable, instructions were given to refer to specific definitions presented.

Network management practices. Indicators related to service interruptions, such as direct indicators of frequency, durations, and related indexes are commonly used in the industry to assess the availability of the service. See, for instance, the System Average Interruption Frequency Index (SAIFI), System Average Interruption Duration Index (SAIDI), and Customer Average Interruption Duration Index (CAIDI). Still, the accuracy and comparability of these indicators will depend on the implementation and configuration of network management automated recording systems (such as fault incidence recording systems or supervisory control and data acquisition [SCADA]/energy management system [EMS] with capabilities to perform these function).

References

AF-MERCADOS EMI. 2013. "Final Report on Proposal of Key Performance Indicators in the Energy Sector in Serbia." Unpublished report, Madrid, Spain.

Andres, Luis A., J. Schwartz, and J. L. Guasch. 2012. "Uncovering the Drivers of Utility Performance: Lessons from Latin America and the Caribbean on the Role of the Private Sector, Regulation, and Governance in the Power, Water, and Telecommunication Sectors." World Bank, Washington, DC.

Banerjee, Sudeshna Ghosh, A. Moreno, J. Sinton, T. Primiani, and J. Seong. 2016. "Regulatory Indicators for Sustainable Energy: A Global Scorecard for Policy Makers." World Bank, Washington, DC.

Eberhard, A., V. F oster, C. Briceño-Garmendia, F. Ouedraogo, D. Camos, and M. Shkaratan. 2008. "Underpowered: The State of the Power Sector in Sub-Saharan Africa." Africa Infrastructure Country Diagnostic (AICD), summary of background paper 6, World Bank, Washington, DC.

Hesmondhalgh, Serena, W. Zarakas, T. Brown. 2012. "Approaches to Setting Electric Distribution Reliability Standards and Outcomes." The Brattle Group.

Trimble, C., M. Kojima, I. P. Arroyo, and F. Mohammadzadeh. 2016. "Financial Viability of Electricity Sectors in Sub-Saharan Africa: Quasi-Fiscal Deficits and Hidden Costs." Policy Research Working Paper 7788, World Bank, Washington, DC.

Utilities Considered and Their Basic Characteristics

The MENA Electricity Database (MED) covers 67 utilities in 14 MENA economies. Table B.1 provides basic power sector characteristics of the 14 MENA economies of interest, including its size, market structure, and key stakeholders. Table B.2 lists the 67 MENA utilities included in the MED. Table B.3 presents the five categories of institutional and contextual characteristics of utilities used in chapter 5 to assess the drivers of performance: type, size, ownership, economy income level, and presence of separate regulatory agency. Finally, table B.4 provides a list of the utilities and their respective economies for which observations were available and used in the non-MENA analysis of this study. Non-MENA data used in the study were obtained from 181 economies. Owing to the lack of a time series for the non-MENA data, the latest available year was used from the range 2005 to 2016. Because data used in the MENA analysis were limited to the period 2009 to 2014, non-MENA observations from 2005 to 2008 were associated with the year 2009.

Table B.1 Summary of the Electricity Sector for 14 MENA Economies, 2013

Country or economy	Installed capacity (MW)	Generation	Distribution	Sector regulator?	Vertical integration?	Comments
Algeria	12,949	Société Algérienne de Production de l'Electricité (SPE)	Société de Distribution de l'Electricité et du Gaz d'Alger (SDA), Société de Distribution de l'Electricité et du Gaz du Centre (SDC), Société de Distribution de l'Electricité et du Gaz de l'Est (SDE), and Société de Distribution de l'Electricité et du Gaz de l'Ouest (SDO)	Yes (Commission de Régulation de l'Electricité et du Gaz, CREG)	Single buyer, unbundled	Société Nationale de l'Electricité et du Gaz (SONELGAZ) is a holding company with SPE, transmission assets, ISO, and four distribution companies. For the purpose of this study, SONELGAZ is considered as a VIU in Algeria because of data constraints.
Bahrain	3,934	BOOT model, with several privately owned facilities	Electricity and Water Authority (EWA) (part of Ministry of Electricity and Water)	No	Single buyer, partially unbundled	Single buyer. EWA is considered as a VIU.
Djibouti	100	Electricité de Djibouti (EDD)	EDD	No	Yes	One of only two countries (with the Republic of Yemen) in which access (53%) is an issue.
Egypt, Arab Rep.	29,312	Six public (regional) and nine private (three BOOT and six IPPs)	Nine regional: Alexandria, South Cairo, North Cairo, El-Behera, South Delta, North Delta, Upper Egypt, Middle Egypt, Canal	Yes (Egyptian Electric Utility and Consumer Protection Regulatory Agency, EgyptERA)	Single buyer, unbundled	Egyptian Electricity Holding Company (EEHC) and its affiliates are responsible for generation, transmission, and distribution. EEHC is a single buyer with elements of a monopoly. For the purpose of this study, we consider the six public GUs and the nine regional DUs, and the Egyptian Electricity Transmission Company (EETC) (a TU).

table continues next page

Table B.1 Summary of the Electricity Sector for 14 MENA Economies, 2013 *(continued)*

Country or economy	Installed capacity (MW)	Generation	Distribution	Sector regulator?	Vertical integration?	Comments
Iraq	19,354	Electric Energy Production (EEP)	Electric Energy Distribution (EED)	No	Yes	EEP, EED, and the Electric Energy Transmission are three divisions within the Ministry of Electricity.
Jordan	3,452	Samra Electric Power Generating Company (SEPCO), state owned (25%); Central Electricity Generating Company (CEGCO), private (56%); Amman East Power Plant (AES PSC), private (16%)	Three private: Jordan Electric Power Company (JEPCO), Electricity Distribution Company (EDCO), Irbid District Electricity Company (IDECO)	Yes (Electricity Regulatory Commission, ERC)	Single buyer, unbundled	Transmission activities are under the National Electricity Power Company (NEPCO).
Lebanon	2,313	Electricité du Liban (EdL) and IPPs	EdL and three DSPs with two-year "concession" contracts	No	Yes	
Morocco	6,677	Office National de l'Electricité et l'Eau Potable (ONEE) and several IPPs	ONEE (50% customers) and four concessions—Lyonnaise des Eaux de Casablanca (LYDEC) for Casablanca, REDAL for Rabat, AMENDIS-TA for Tanger, AMENDIS-TE for Tetouan—and seven municipal multiservice distribution utilities	No	Single buyer, partially unbundled	There are some private players in generation and distribution.
Oman	4,938	Some public and several private actors: United Power Company (UPC), Sembcorp Salalah Power Company (SSPWC), Al Rusail Power Plant (ARPP), Barka Power and Desalination Plant (BPDP), Al Kamil Power Plant (AKPP), Phoenix Power Company (PPC), ACWA Power Barka (APBS), Shoar Power Plant (SPP).	Three state-owned distribution and supply companies	Yes (Authority for Electricity Regulation, AER)	Single buyer, unbundled	Oman Electricity Transmission Company (OTEC) is responsible for transmission of electricity and Oman Power and Water Procurement Company (OPWPC) is the single buyer of electricity and water.

table continues next page

Table B.1 Summary of the Electricity Sector for 14 MENA Economies, 2013 *(continued)*

Country or economy	Installed capacity (MW)	Generation	Distribution	Sector regulator?	Vertical integration?	Comments
Qatar	8,756	Private sector and IWPPs	Qatar General Electricity and Water Corporation (KAHRAMAA)	No	Single buyer, unbundled	
Saudi Arabia	53,588	Saudi Electricity Company (SEC) and several private: Saline Water Conversion Corporation (SWCC), Saudi Aramco, Tihamah, Power and Utility Company for Jubail and Yanbu (MARAFIQ), Water and Electricity LLC. (WEC) including several large industrial firms	SEC	Yes (Electricity and Cogeneration Regulatory Authority, ECRA)	Yes	SEC is a vertically integrated monopoly. Power for desalination is an important player. Saudi Arabia plans to unbundle SEC soon.
Tunisia	4,095	Société Tunisienne de l'Electricité et du Gaz (STEG) and two IPPs	STEG	No	Yes	
West Bank	125	Electricity supply almost entirely dependent on the Israel Electric Corporation (IEC) and Gaza IPP (2x70 MW)	Five: Northern Electricity Distribution Company (NEDCO) for North West Bank, Southern Electricity Company (SELCO) for South West Bank, Hebron Electricity Corporation (HEPCO) for Hebron, Jerusalem District Electricity Company (JEDCO) for Jerusalem, Gaza Electricity Distribution Company (GEDCO) for Gaza	Yes (Palestinian Electricity Regulatory Council, PERC)	Single buyer, unbundled	
Yemen, Rep.	1,520	Public Electricity Corporation (PEC)	PEC	No	Yes	PEC is vertically integrated. Yemen, Rep., is one of only two countries of the sample in which access (48%) is an issue, along with Djibouti.

Source: World Bank calculations.

Note: BOOT = build-own-operate-transfer; DSP = distribution service provider; DU = distribution utility; GU = generation utility; IPP = independent power producer; ISO = independent system operator; IWPP = independent water and power producer; MENA = Middle East and North Africa; MW = megawatts; TU = transmission utility; VIU = vertically integrated utility.

Table B.2 Names and Abbreviations of MENA Utilities

Country or economy	Utility name	Abbreviation
Algeria	Société Nationale de l'Électricité et du Gaz	SONELGAZ
Bahrain	Electricity and Water Authority	EWA
Djibouti	Électricité de Djibouti	EdD
Egypt, Arab Rep.	Alexandria Electricity Distribution Company	AEDC
	Cairo Electricity Production Company	CEPC
	Canal Electricity Distribution Company	CEDC
	East Delta Electricity Production Company	EDEPC
	Egyptian Electricity Transmission Company	EETC
	El-Behera Electricity Distribution Company	EEDC
	Middle Delta Electricity Production Company	MDEPC
	Middle Egypt Electricity Distribution Company	MEEDC
	North Cairo Electricity Distribution Company	NCEDC
	North Delta Electricity Distribution Company	NDEDC
	South Cairo Electricity Distribution Company	SCEDC
	South Delta Electricity Distribution Company	SDEDC
	Upper Egypt Electricity Distribution Company	UEEDC
	Upper Egypt Electricity Production Company	UEEPC
	West Delta Electricity Production Company	WDEPC
Iraq	Ministry of Electricity	MOE
Jordan	AES Levant Holding BV Jordan PSC	AES Levant
	Amman East Power Plant	AES PSC
	Amman-Asia Electric Generating Company	AAEPC
	Central Electricity Generating Company	CEGCO
	Electricity Distribution Company	EDCO
	Irbid District Electricity Company	IDECO
	Jordan Electric Power Company	JEPCO
	National Electric Power Company	NEPCO
	Qatrana Electric Power Company	QEPCO
	Samra Electric Power Generating Company	SEPCO
Lebanon	Électricité du Liban	EdL
Morocco	AMENDIS Tanger	AMENDIS TANGER
	AMENDIS Tetouan	AMENDIS TETOUAN
	Lyonnaise des Eaux de Casablanca	LYDEC
	Office National de l'Électricité et de l'Eau Potable	ONEE
	RADEEL	RADEEL
	REDAL Rabat	REDAL
	Régie Autonome de Distribution d'Eau d'Électricité et d'Assainissement liquide de la Province de Kenitra	RAK
	Régie Autonome de Distribution d'Eau et d'Électricité de Marrakech	RADEEMA
	Régie Autonome de Distribution d'Eau et d'Électricité de Meknès	RADEM
	Régie Autonome de Distribution d'Eau, d'Électricité et d'Assainissement liquide des Provinces d'El Jadida et de Sidi Bennour	RADEEJ
	Régie Autonome Intercommunale de Distribution d'Eau et d'Électricité de Safi	RADEES
	Régie Autonomie Intercommunale de Distribution d'Eau et d'Électricité de Fès	RADEEF

table continues next page

Table B.2 Names and Abbreviations of MENA Utilities *(continued)*

Country or economy	Utility name	Abbreviation
Oman	ACWA Power Barka	APBS
	Al Batinah Power Company	ABPC
	Al Suwadi Power Company	ASPC
	Al-Ghubra Power and Desalination Company	GPDCO
	Al-Kamil Power Plant	AKPP
	Al-Rusail Power Plant	ARPP
	Barka Power and Desalination Plant	BPDP
	Dhofar Power Company	DPC
	Majan Electricity Company	MJEC
	Mazoon Electricity Distribution Company	MZEC
	Muscat Electricity Distribution Company	MEDC
	Oman Electricity Transmission Company	OETC
	Phoenix Power Company	PPC
	Rural Areas Electricity Company	RAECO
	Sembcorp Salalah Power and Water Company	SSPWC
	Sohar Power Plant	SPP
	United Power Company	UPC
	Wadi Al-Jizzi Power Company	WAJPCO
Qatar	Qatar General Electricity and Water Corporation	KAHRAMAA
Saudi Arabia	Saudi Electricity Company	SEC
Tunisia	Société Tunisienne de l'Électricité et du Gaz	STEG
West Bank	Jerusalem District Electricity Company	JDECO
	Northern Electricity Distribution Company	NEDCO
	Tubas District Electricity Company	TUBAS
Yemen, Rep.	Public Electricity Corporation	PEC

Source: MENA Electricity Database.
Note: MENA = Middle East and North Africa.

Table B.3 Characteristics of MENA Utilities

Country or economy	Utility name	Type	Size	Ownership	Country income level	Presence of separate regulatory agency
Algeria	Société Nationale de l'Électricité et du Gaz	VIU	Big	Public	Upper middle	Yes
Bahrain	Electricity and Water Authority	VIU	Medium	Public	High	No
Djibouti	Électricité de Djibouti	VIU	Small	Public	Lower middle	No
Egypt, Arab Rep.	Alexandria Electricity Distribution Company	DU	Big	Public	Lower middle	Yes
	Cairo Electricity Production Company	GU	Big	Public	Lower middle	Yes
	Canal Electricity Distribution Company	DU	Big	Public	Lower middle	Yes
	East Delta Electricity Production Company	GU	Big	Public	Lower middle	Yes

table continues next page

Table B.3 Characteristics of MENA Utilities *(continued)*

Country or economy	Utility name	Type	Size	Ownership	Economy income level	Presence of separate regulatory agency
	Egyptian Electricity Transmission Company	TU	Big	Public	Lower middle	Yes
	El-Behera Electricity Distribution Company	DU	Medium	Public	Lower middle	Yes
	Middle Delta Electricity Production Company	GU	Big	Public	Lower middle	Yes
	Middle Egypt Electricity Distribution Company	DU	Big	Public	Lower middle	Yes
	North Cairo Electricity Distribution Company	DU	Big	Public	Lower middle	Yes
	North Delta Electricity Distribution Company	DU	Big	Public	Lower middle	Yes
	South Cairo Electricity Distribution Company	DU	Big	Public	Lower middle	Yes
	South Delta Electricity Distribution Company	DU	Big	Public	Lower middle	Yes
	Upper Egypt Electricity Distribution Company	DU	Big	Public	Lower middle	Yes
	Upper Egypt Electricity Production Company	GU	Big	Public	Lower middle	Yes
	West Delta Electricity Production Company	GU	Big	Public	Lower middle	Yes
Iraq	Ministry of Electricity	VIU	Big	Public	Upper middle	No
Jordan	AES Levant Holding BV Jordan psc	GU	Small	Private	Upper middle	Yes
	Amman-Asia Electric Generating Company	GU	Medium	Private	Upper middle	Yes
	Amman East Power Plant	GU	Small	Private	Upper middle	Yes
	Central Electricity Generating Company	GU	Big	Private	Upper middle	Yes
	Electricity Distribution Company	DU	Small	Private	Upper middle	Yes
	Irbid District Electricity Company	DU	Medium	Private	Upper middle	Yes
	Jordan Electric Power Company	DU	Medium	Private	Upper middle	Yes
	National Electric Power Company	TU	Big	Public	Upper middle	Yes
	Qatrana Electric Power Company	GU	Small	Private	Upper middle	Yes
	Samra Electric Power Generating Company	GU	Big	Public	Upper middle	Yes
Lebanon	Électricité du Liban	VIU	Medium	Public	Upper middle	No
Morocco	AMENDIS Tanger	DU	Medium	Private	Lower middle	No
	AMENDIS Tetouan	DU	Small	Private	Lower middle	No

table continues next page

Table B.3 Characteristics of MENA Utilities *(continued)*

Country or economy	Utility name	Type	Size	Ownership	Economy income level	Presence of separate regulatory agency
	Lyonnaise des Eaux de Casablanca	DU	Medium	Private	Lower middle	No
	Office National de l'Électricité et de l'Eau Potable	VIU	Big	Public	Lower middle	No
	RADEEL	DU	Small	Public	Lower middle	No
	REDAL Rabat	DU	Medium	Private	Lower middle	No
	Régie Autonome de Distribution d'Eau d'Électricité et d'Assainissement liquide de la province de Kenitra	DU	Small	Public	Lower middle	No
	Régie Autonome de Distribution d'Eau et d'Électricité de Marrakech	DU	Medium	Public	Lower middle	No
	Régie Autonome de Distribution d'Eau et d'Électricité de Meknès	DU	Small	Public	Lower middle	No
	Régie Autonome de Distribution d'Eau, d'Électricité et d'Assainissement liquide des Provinces d'El Jadida et de Sidi Bennour	DU	Small	Public	Lower middle	No
	Régie Autonomie Intercommunale de Distribution d'Eau et d'Électricité de Fès	DU	Medium	Public	Lower middle	No
	Régie Autonome Intercommunale de Distribution d'Eau et d'Électricité de Safi	DU	Small	Public	Lower middle	No
Oman	ACWA Power Barka	GU	Small	Private	High	Yes
	Al Batinah Power Company	GU	Medium	Public	High	Yes
	Al Suwadi Power Company	GU	Medium	Public	High	Yes
	Al-Ghubra Power and Desalination Company	GU	Small	Public	High	Yes
	Al-Kamil Power Plant	GU	Small	Private	High	Yes
	Al-Rusail Power Plant	GU	Medium	Private	High	Yes
	Barka Power and Desalination Plant	GU	Medium	Private	High	Yes

table continues next page

Table B.3 Characteristics of MENA Utilities (continued)

Country or economy	Utility name	Type	Size	Ownership	Economy income level	Presence of separate regulatory agency
	Dhofar Power Company	VIU	Small	Private	High	Yes
	Majan Electricity Company	DU	Small	Public	High	Yes
	Mazoon Electricity Distribution Company	DU	Medium	Public	High	Yes
	Muscat Electricity Distribution Company	DU	Medium	Public	High	Yes
	Oman Electricity Transmission Company	TU	Big	Public	High	Yes
	Phoenix Power Company	GU	Big	Private	High	Yes
	Rural Areas Electricity Company	VIU	Small	Public	High	Yes
	Sembcorp Salalah Power and Water Company	GU	Small	Private	High	Yes
	Sohar Power Plant	GU	Medium	Private	High	Yes
	United Power Company	GU	Small	Private	High	Yes
	Wadi Al-Jizzi Power Company	GU	Small	Public	High	Yes
Qatar	Qatar General Electricity and Water Corporation	VIU	Medium	Public	High	No
Saudi Arabia	Saudi Electricity Company	VIU	Big	Public	High	Yes
Tunisia	Société Tunisienne de l'Électricité et du Gaz	VIU	Big	Public	Upper middle	Yes
West Bank	Jerusalem District Electricity Company	DU	Small	Public	Lower middle	Yes
	Northern Electricity Distribution Company	DU	Small	Public	Lower middle	Yes
	Tubas District Electricity Company	DU	Small	Private	Lower middle	Yes
Yemen, Rep.	Public Electricity Corporation	VIU	Medium	Public	Lower middle	Yes

Source: MENA Electricity Database.
Note: DU = distribution utility; GU = generation utility; MENA = Middle East and North Africa; TU = transmission utility; VIU = vertically integrated utility.

Table B.4 Names and Abbreviations of Non-MENA Utilities

Economy	Utility name	Abbreviation
Angola	Empresa de Electricidade de Luanda	ANG-EDEL
Antigua and Barbuda	Antigua Public Utilities Authority	ANT-APUA
Argentina	Empresa Distribuidora de Energía Atlántica S.A.	ARG-EDE
	Empresa Distribuidora de Electricidad de Mendoza S.A.	ARG-EDMSA
	Empresa Distribuidora de Electricidad de Santiago del Estero S.A.	ARG-EDSTE
	Empresa Distribuidora de Electricidad de San Luis S.A.	ARG-EDSL
	Empresa Distribuidora Norte S.A.	ARG-EDNR
	Empresa Distribuidora Sur S.A.	ARG-EDSR
Australia	AGL Electricity Ltd.	AUS-AGL
Belize	Belize Electricity Limited	BEL-BECOL
Benin	Electricité de Benin	BEN-CEB
Bolivia	Cooperativa Rural de Electrificación Ltda.	BOL-CRE
	Empresa de Luz y Fuerza Eléctrica Cochabamba	BOL-ELFEC
Botswana	Botswana Power Corporation	BOT-BPC
Burkina Faso	Société Nationale d'Electricité du Burkina	BUR-SONABEL
Brazil	AES SUL Distribuidora Gaúcha de Energia S.A.	BRA-AESUL
	Ampla Energia e Serviços S.A.	BRA-AMPLA
	Caiuá Serviços de Eletricidade S.A.	BRA-CAI
	Centrais Elétricas de Santa Catarina S. A.	BRA-CELSC
	Centrais Elétricas do Pará S.A.	BRA-CELPA
	Centrais Elétricas Matogrossenses S.A.	BRA-CMT
	CMG	BRA-CMG
	Companhia Campolarguense de Energia	BRA-COC
	Companhia de Eletricidade do Amapá	BRA-CEA
	Companhia de Eletricidade do Amapá	BRA-CEAM
	Companhia de Eletricidade do Estado da Bahia	BRA-COELB
	Companhia de Energia Elétrica do Estado do Tocantins	BRA-CELT
	Companhia Energética de Alagoas	BRA-CEAL
	Companhia Energética de Brasília	BRA-CEB
	Companhia Energética de Goiás	BRA-CELG
	Companhia Energética de Minas Gerais S.A.	BRA-CEMIG
	Companhia Energética de Pernambuco	BRA-CELPE
	Companhia Energética de Roraima	BRA-CER
	Companhia Energética de São Paulo	BRA-CPSA
	Companhia Energética do Ceará	BRA-COELC
	Companhia Energética do Maranhão	BRA-CMR
	Companhia Energética do Rio Grande do Norte	BRA-COS
	Companhia Estadual de Energia Elétrica	BRA-CEEE
	Companhia Força e Luz do Oeste	BRA-CFLO
	Companhia Hidroelétrica São Patrício	BRA-CHSP
	Companhia Jaguari de Energi	BRA-CJE
	Companhia Luz e Força Mococa	BRA-MOCCA
	Companhia Luz e Força Santa Cruz	BRA-SANT
	Companhia Nacional de Energia Elétrica	BRA-CNEE

table continues next page

Table B.4 Names and Abbreviations of Non-MENA Utilities (continued)

Economy	Utility name	Abbreviation
	Companhia Paranaense de Energia	BRA-COP
	Companhia Paulista de Força e Luz	BRA-PAUL
	Companhia Piratininga de Força e Luz	BRA-PIRA
	COO	BRA-COO
	CPE	BRA-CPE
	CRN	BRA-CRN
	CSPE	BRA-CSPE
	Departamento Municipal de Energia de Ijuí	BRA-DEM
	DME Distribuição S.A.	BRA-EBO
	Electricidade de São Paulo S.A.	BRA-ELETROPAULO
	Empresa Bandeirante de Energia	BRA-BAND
	Empresa Elétrica Bragantina S.A.	BRA-EEB
	Energisa Borborema	BRA-BOA
	Energisa Borborema S.A.	BRA-DME
	Metropolitana Eletricidade de São Paulo S.A.	BRA-AES
Cabo Verde	Electra	CAB-ELECTRA
Cameroon	Cameroon Electricity Corporation	CAM-CEC
Chad	Société Tchadienne d'Eau et d'Electricité	CHA-STEE
Chile	Compañía General de Electricidad Distribución S.A.	CHL-CGED
	Chilectra S. A.	CHL-CHIL
	Compañía Nacional de Fuerza Eléctrica S.A.	CHL-CON
Colombia	Centrales Eléctricas de Nariño S.A. E.S.P.	COL-CEDENAR
	Centrales Eléctricas del Cauca S.A. E.S.P.	COL-CEDELCA
	Centrales Eléctricas del Norte de Santander S.A. E.S.P.	COL-CENS
	CHC Energía	COL-CHC
	COD	COL-COD
	Codensa S.A. E.S.P.	COL-CODENSA
	Electrificadora de Santander	COL-ESSA
	Electrohuila S.A. E.S.P	COL-ELECTROHUILA
	Empresa Eléctrica de Colina Ltda.	COL-EEC
Comoros	Electricite et Eaux des Comores	COM-EEDC
Congo, Dem. Rep.	Société Nationale d'Electricité	DRC-SNEL
Costa Rica	Compañia Nacional de Fuerza y Luz	COS-CNFL
	Instituto Costarricense de Electricidad	COS-ICE
Croatia	Distribucijskog sustava	CRO-HEP-ODS
	Hrvatska elektroprivreda	CRO-HEP
Denmark	Dong Energy	DEN-DONG
Dominica	Dominica Electricity Services Limited	DOM-DOMLEC
Dominican Republic	Empresa Distribuidora de Electricidad del Este	DOM-EDESTE
	Empresa Distribuidora de Electricidad del Norte	DOM-EDENOTRE
	Empresa Distribuidora de Electricidad del Sur	DOM-EDESUR
Ecuador	Corporación para la Administración Temporal Eléctrica de Guayaquil—Distribución y Comercialización	ECU-CATEG-D/EMELEC
	Empresa Eléctrica Ambato S.A.	ECU-AMBATO

table continues next page

Table B.4 Names and Abbreviations of Non-MENA Utilities *(continued)*

Economy	Utility name	Abbreviation
	Empresa Eléctrica Azogues S.A.	ECU-AZOGUES
	Empresa Eléctrica Bolivar S.A.	ECU-BOLIVAR
	Empresa Eléctrica Cotopaxi S.A.	ECU-COTOPAXI
	Empresa Eléctrica El Oro S.A.	ECU-ELORO
	Empresa Eléctrica Esmeraldas S.A.	ECU-ESMERALDAS
	Empresa Eléctrica Galapagos S.A.	ECU-GALAPAGOS
	Empresa Eléctrica Guayas Los Ríos S.A.	ECU-GUAYAS-LOSRÍOS
	Empresa Eléctrica Los Ríos S.A.	ECU-LOS RIOS
	Empresa Eléctrica Manabí S.A.	ECU-MANABÍ
	Empresa Eléctrica Milagro S.A.	ECU-MILAGRO
	Empresa Eléctrica Notre S.A.	ECU-NORTE
	Empresa Eléctrica Quito S.A.	ECU-QUITO
	Empresa Eléctrica Regional Centro Sur S.A.	ECU-CENTROSUR
	Empresa Eléctrica Regional Sur S.A.	ECU-SUR
	Empresa Eléctrica Riobamba S.A.	ECU-RIOBAMBA
	Empresa Eléctrica Santa Elena S.A.	ECU-STA.ELENA
	Empresa Eléctrica Santo Domingo S.A.	ECU-STO.DOMINGO
	Empresa Eléctrica Sucumbíos S.A.	ECU-SUCUMBÍOS
El Salvador	AES-El Salvador	ELS-AES
	CAESS	ELS-CAE
	CLESA	ELS-CLSA
	Distribuidora Eléctrica de Usulután (DEUSEM)	ELS-DEU
	Duke Energy El Salvador Co.	ELS-DEL
	La Empresa Eléctrica de Oriente	ELS-EEO
Ethiopia	Ethiopian Electric Power	ETH-EEP
France	Réseau de Transport d'Électricité	FRA-RTE
Ghana	Electricity Company of Ghana Ltd.	GHA-ECG
Grenada	Grenada Electricity Services Ltd.	GRE-GRENLEC
Guinea	Electricité de Guinée	GUI-EDG
Guinea-Bissau	National Electricity and Water Corporation	GUB-EAGB
Honduras	Empresa Nacional de Energía Eléctrica	HON-ENEE
India	Ajmer Vidyut Vitran Nigam Ltd.	IND-AVVNL
	BSES Yamuna Power Limited	IND-BYPL
	Dakshin Gujarat Vij Company Ltd.	IND-DGVCL
	Gujarat Energy Transmission Corporation Limited	IND-GETCO
	Gujarat State Electricity Corporation Limited	IND-GSECL
	Jaipur Vidyut Vitran Nigam Ltd.	IND-JVVNL
	Jodhpur Vidyut Vitran Nigam Ltd.	IND-JdVVNL
	Madhya Gujarat Vij Company Limited	IND-MGVCL
	North Delhi Power Limited	IND-NDPL
	Paschim Gujarat Vij Company Ltd.	IND-PGVCL
	Rajasthan Rajya Vidyut Prasaran Nigam Ltd.	IND-RVPNL
	Rajasthan Rajya Vidyut Utpadan Nigam Ltd.	IND-RVUNL
Israel	Israel Electric Corporation	ISR-IEC

table continues next page

Table B.4 Names and Abbreviations of Non-MENA Utilities *(continued)*

Economy	Utility name	Abbreviation
Jamaica	Jamaica Public Service Company Ltd.	JAM-JPSco
Kenya	Kenya Power	KEN-KPLC
Lesotho	Lesotho Electricity Company	LES-LEC
Malawi	Electricity Supply Corporation of Malawi Ltd.	MLW-ESCOM
Mauritius	Central Electricity Board	MAU-CEB
Mexico	Comisión Federal de Electricidad	MEX-CFE
	Luz y Fuerza del Centro	MEX-LyFC
Mozambique	Electricidade de Moçambique	MOZ-EDM
Namibia	Nampower	NAM-NAMPOWER
Paraguay	Administración Nacional de Electricidad	PAR-ANDE
Peru	Consorcio Eléctrico de Villacurí S.A.C.	PER-COELVISAC
	EDELNOR	PER-EDELNOR
	Electro Centro S.A.	PER-ELC
	Electro Nor Oeste S.A.	PER-ENOSA
	Electro Norte S.A.	PER-ENSA
	Electro Oriente S. A	PER-ELOR
	Electro Pangoa S.A.	PER-Pangoa
	Electro Puno S.A.A.	PER-ELPUNO
	Electro Sur Este S.A.	PER-ELSE
	Electro Sur Medio S.A.A.	PER-ELSM
	Electro Sur S.A.	PER-ELS
	Electro Tocache	PER-Tocache
	Electro Ucayali S.A.	PER-ELU
	Electronorte Medio S.A.-Hidradina S.A.	PER-ELECTRONORTEMEDIO
	Empresa de Distribución Eléctrica Cañete S.A.	PER-EDECAÑETE
	Empresa de Servicios Eléctricos Municipales de Paramonga S. A.	PER-EMSEMSA
	Empresa Municipal de Servicios Eléctricos Utcubamba	PER-EMSEU
	Luz del Sur	PER-LUZ del Sur
	Servicios Eléctricos Rioja	PER-SERSA
	Sociedad Electrica del Sur Oeste S.A.	PER-SEAL
Portugal	EDP Distribuição	POR-EDP-DIS
	Energias de Portugal	POR-EDP
Rwanda	Rwanda Energy Group	RWA-REG
Senegal	Société National d'Éléctricité du Sénégal	SEN-SENELEC
Sierra Leone	National Power Authority	SIE-NPA
South Africa	The Electricity Supply Commission	ZAF-ESKOM
Spain	Red Eléctrica de España	SPA-REE
Sri Lanka	Ceylon Electricity Board	SRI-CEB
St. Kitts and Nevis	St. Kitts Electricity Department	STK-SED
St. Lucia	St. Lucia Electricity Services Limited	STL-LUCELEC
Swaziland	Swaziland Electricity Company	SWA-SEC
Tanzania	Tanzania Electric Supply Company Limited	TAN-TANESCO
Togo	Compagnie Energie Electrique du Togo	TOG-CEET

table continues next page

Table B.4 Names and Abbreviations of Non-MENA Utilities *(continued)*

Economy	Utility name	Abbreviation
Uganda	Uganda Electricity Board	UGA-UEB
Uruguay	Administración Nacional de Usinas y Trasmisiones Eléctricas	URU-UTE
United States	Consolidated Edison Inc.	USA-ConEdison
	Duke Energy	USA-Duke
Venezuela, RB	La Electricidad de Caracas S.A.	EDC-AES
Vietnam	VietNam Electricity	VIET-EVN
Zambia	Zambia Electricity Supply Corporation Limited	ZAM-ZESCO

Source: MENA Electricity Database.

Quasi-Fiscal Deficit: Hypothesis and Methodology

Data Sources and Definitions of Key Variables

Table C.1 presents the sources of data that are used for each variable of the economy-level quasi-fiscal deficit (QFD) calculations for the 14 economies considered.

As far as the utility-level QFD is concerned, the MENA Electricity Database (MED) was used to fill in all variables except for the gross domestic product, for which the World Development Indicators were used. However, in some cases alternate sources to the MED were used. These exceptions are listed in table C.2.

In order to compute the components of the economy-level QFD, a series of assumptions and approximations were used in some cases, as described in table C.3.

Table C.1 Sources of Data Used for the Economy-Level QFD Calculations

Country or economy	Qe: end-user consumption (kWh)	Tc: cost-recovery tariff	Te: avg. End-user tariff	Lm: technical loss rates[a]	Number of customers (connections)	Number of employees (FTE)	Cost of labor[c]	Rct: collection rates	GDP
Algeria	MED	Calculations (WDI; ESMAP META Model; Lazard's LCOE Analysis, 2014)	Arab Union of Electricity (2014), Electricity Tariff in the Arab Countries	WDI	MED	MED	MED	Online	WDI
Bahrain	WDI				MED	Online	MED	MED	
Djibouti	MED			Online[b]	MED	MED	MED	MED	
Egypt, Arab Rep.	WDI			WDI	EEHC Annual Report 2014	EEHC Annual Report 2014	Estimation	MED (average)	
Iraq	MED			WDI	MED	Online[d]	MED	Online (World Bank)[e]	
Jordan	WDI			WDI	NEPCO Annual Report 2013	NEPCO Annual Report 2013	Estimation MED	MED (average)	
Lebanon	WDI			WDI	MED	MED	MED	Online[f]	
Morocco	WDI			WDI			Estimation MED	ONEE contact	
Oman	WDI			WDI	AER Annual Report 2013	AER Annual Report 2013	Estimation MED	Estimated	

table continues next page

Table C.1 Sources of Data Used for the Economy-Level QFD Calculations *(continued)*

Country or economy	Qe: End-user consumption (kWh)	Tc: Cost-recovery tariff	Te: Avg. End-user tariff	Lm: Technical loss rates[a]	Number of customers (connections)	Number of employees (FTE)	Cost of labor[b]	Rct: Collection Rates	GDP
Qatar	WDI			WDI	KAHRAMAA Sustainability Report 2013	KAHRAMAA Sustainability Report 2013	KAHRAMAA Annual Report 2014	MED	
Saudi Arabia	WDI			WDI	MED	MED	MED	SEC statistics 2000 to 2014	
Tunisia	WDI			WDI		MED		Data from utility	
West Bank	MED[g]			MED (average)		MED		MED (average)	
Yemen, Rep.	WDI			WDI		MED[h]	Estimated	MED	

Source: World Bank calculations.

Note: AER = Authority for Electricity Regulation; EEHC = Egyptian Electricity Holding Company; ESMAP = Energy Sector Management Assistance Program; FTE = full-time equivalent employee; GDP = gross domestic product; KAHRAMAA = Qatar General Electricity and Water Corporation; kWh = kilowatt-hours; LCOE = levelized cost of electricity; MED = MENA Electricity Database; META = Model for Electricity Technology Assessment; NEPCO = National Electric Power Company; ONEE = Office National de l'Électricité et de l'Eau Potable; SEC = Saudi Electricity Company; WDI = World Development Indicators.

a. WDI technical losses (distribution and transmission losses).

b. EUEI (European Union Energy Initiative) 2013.

c. Refer to appendix tables C.7 onward for calculation details.

d. Iraq Energy Institute 2015.

e. World Bank 2016a.

f. Lebanon Ministry of Environment and UNDP.

g. Calculated as the sum of energy volume billed (from MED) for the three distribution utilities in the West Bank (TUBAS, JDECO, and NEDCO).

h. Used 2012 value in the case of the Republic of Yemen owing to lack of data for 2013.

Table C.2 List of Alternate Sources for the Utility-Level QFD

Utility	Indicator	Source
Algeria: Sociètè Nationale de l'Electricité et du Gaz (SONELGAZ)	Share of loss (%)	WDI.
	Bill collection rate	L'Algérie profonde/Ouest. n.d. "Plus de 184 millions de DA de pertes pour la Sonelgaz." http://www.liberte-algerie.com /ouest/plus-de-184-millions-de-da-de-pertes-pour-la -sonelgaz-228183/print/1.
Bahrain: Electricity and Water Authority (EWA)	Share of loss (%)	WDI.
Djibouti: Electricité de Djibouti (EDD)	Total electricity billed	EUEI (European Union Energy Initiative). 2013. *Country Power Market Brief*: Djibouti. Africa-EU Energy Partnership. http:// www.euei-pdf.org/sites/default/files/field_publication_file / AEEP_Djibouti_Country_market_brief_EN.pdf.
	Bill collection rate	MED: 2013 value used for 2011 calculations for lack of data.
Iraq: Ministry of Electricity (MoE)	Bill collection rate	World Bank. 2016a. "eC2: Electricity Services Restoration and Operations Efficiency." ToR for World Bank Assignment Title: 1223732—IRAQ, Netherlands for the World Bank, July 24. https://nl4worldbank.org/2016/07/14/ec2electricity-services -restoration-and-operations-efficiency.
	Number of employees	Approximation for number of employees in the Ministry of Electricity.
Lebanon: Electricité du Liban (EdL)	Number of new customers	Estimated from WDI population growth figures for 2011.
	Length of transmission network	AUE (Arab Union of Electricity). 2013. *Statistical Bulletin 2013*. Amman, Jordan: AUE.
	Share of loss (%)	WDI.
Saudi Arabia: Saudi Electricity Company (SEC)	Bill collection rate	Calculated from SEC document "Statistics 2000 to 2014." See appendix table C.4 for methodology.
Tunisia: Société Tunisienne de l'Electricité et du Gaz (STEG)	Bill collection rate	Value obtained from STEG.
West Bank: Northern Electricity Distribution Company (NEDCO)	Energy purchased	Calculated from electricity IEC sold to West Bank in 2013. World Bank. 2014b . *West Bank and Gaza: Assessment and Action Plan to Improve Payment for Electricity Services in the Palestinian Territories: Study on Electricity Sector Contribution to Net Lending.* Report No: ACS9393. Washington, DC: World Bank. http:// documents.worldbank.org/curated/en/120271468317065014 /pdf/ACS93930WP0P1469990Box385388B00OUO090.pdf.
	Total electricity billed	World Bank. 2014b. *West Bank and Gaza—Assessment and Action Plan to Improve Payment for Electricity Services in the Palestinian Territories: Study on Electricity Sector Contribution to Net Lending.* Report No: ACS9393. Washington, DC: World Bank.
West Bank: Tubas District Electricity Company (TUBAS)	Energy purchased	World Bank. 2014b. *West Bank and Gaza—Assessment and Action Plan to Improve Payment for Electricity Services in the Palestinian Territories: Study on Electricity Sector Contribution to Net Lending.* Report No: ACS9393. Washington, DC: World Bank.
Yemen, Rep.: Public Electricity Corporation (PEC)	Length of transmission network	AUE (Arab Union of Electricity). 2013. *Statistical Bulletin 2013*. Amman, Jordan: AUE.
	Bill collection rate	World Bank. 2014a. "YEM Power Ministerial Note." Unpublished paper, Washington, DC. January 3.

Note: IEC = Israel Electric Corporation; MED = MENA Electricity Database; MENA = Middle East and North Africa; WDI = World Development Indicators.

Table C.3 Descriptions and Assumptions of Economy-Level QFD Components

Element	Description and assumptions
(Qe) End-user consumption	Calculated by multiplying the electric power consumption per capita by the total population of the economy for the year 2013.
(Te) Average end-user tariff	Taken to be the average residential tariff for a consumption of 250 kWh/month for the year 2013. Values for all economies were calculated based upon the Arab Union of Electricity's (2014), "Electricity Tariff in the Arab Countries." In the case of Djibouti, calculations were based upon the official tariff document published by the economy.
(Tc) Cost-recovery tariff rate	Not readily available and had to be estimated using the LCOE. The LCOE unit cost of energy per technology type was obtained in $/kWh and then weighted according to the energy mix of each economy. Sources used were WDI for the energy mix information, and an LCOE modeling tool developed by ESMAP[a] for most of the LCOE values. Because the unit cost of fuel and renewables used in the modeling tool did not reflect the current state of energy sources in the MENA region, values from Lazard's LCOE Analysis 2014 were used instead (see also appendix table C.5). These values do not consider the T&D contribution to the unit cost, and for this reason, a factor of ¢ 3.2/kWh was added to ensure that the T&D costs were considered in the calculations.
(Lm) Technical loss rate	The technical loss rate is defined as the electric power transmission and distribution losses (% of output) and was obtained from WDI database. WDI did not include data for West Bank (calculated alternatively as the average of the technical losses of West Bank distribution utilities in the MED) and Djibouti (value obtained as the grid losses from an online source).[b]
(Ln) Normative loss rate	The choice of 5% was done so as to have values of Ln below the region's best-performing economies, namely Bahrain and Qatar with technical loss rates of 5.2% and 6.0%, respectively.
(Rct) Collection rate	The bill collection rate indicates the income effectively collected during the year by the utility in relation to the income billed. In the cases where a single utility existed (a VIU in the case of Algeria, for example), the collection rate of the economy was that of the utility. When more than one utility existed, the average value of the distribution utilities was used (in the case of Egypt, Arab Rep. for example). The collection rate was one of the most challenging indicators to collect from utilities in the MENA region, and when this was not possible, the methodology detailed in appendix C was used with the data presented in table C.19.
(NC) Number of customers (connections)	This figure was easily obtained for economies with a single VIU. For economies with several utilities, the presence of a regulator would allow for an aggregate official figure to be obtained from the regulator's annual report. However, in the case of no regulator present, the sum of individual utility customers was calculated.
(NE) Number of employees	The methodology used to obtain this figure was similar to that of the number of customers. The number of FTE employees was used for all utilities, except in the case of Oman, where the number of total (direct and indirect) employees was used. This is because several utilities in Oman have a very low number of FTE whereas the number of outsourced (or indirect) employees is high. For example, in 2013, the indirect employees in Oman represented 67% of the total of 8,277 employees.

table continues next page

Table C.3 **Descriptions and Assumptions of Economy-Level Quasi-Fiscal Deficit Components** *(continued)*

Element	Description and assumptions
(CL) *Cost of labor*	The cost of labor is defined as the annual cost of personnel directly employed by the utility and was sourced mainly from the financial statements of utilities. However, when this was not available, estimates were made to calculate a unit labor cost per employee, which was then multiplied by the number of employees present in the utilities for which labor cost data were not available. A calculated sum then allowed the economy-level aggregated estimate of the cost of labor to be obtained (see also methodology in appendix tables C.8 to C.16). In the case of the Republic of Yemen, where no labor cost data were available for the VIU (Public Electricity Corporation, PEC), an average unit cost of labor per employee was obtained from average earnings figures from the ILO (see also table C.17 and table C.18).
(413) *Benchmark number of* *customers per* *employee in LICs*	Customer per employee is an indicator of performance with values commonly above 500 in the OECD economies.[b] The value of 413 used in this study was obtained using the same benchmark value for the number of customers per employees in low-income countries as in the AICD methodology.

Source: World Bank calculations, except where noted below.
Note: A compilation of economic costs of more than 50 electricity generation and delivery technologies, META was rolled out to the World Bank Group and selected partners and clients in June 2012. Since then, META has been used in Dominica, the Arab Republic of Egypt, Kosovo, the former Yugoslav Republic of Macedonia, Morocco, and Vietnam as part of the World Bank's engagement in these countries, and by consultants in Haiti and Jamaica. It can be downloaded here: http://esmap.org/META. AICD = Africa Infrastructure Country Diagnostic; ESMAP = Energy Sector Management Assistance Program; FTE = full-time equivalent; ILO = International Labour Organization; kWh = kilowatt-hours; LCOE = levelized cost of electricity; LICs = low-income countries; MED = MENA Electricity Database; MENA = Middle East and North Africa; META = Model for Electricity Technology Assessment; OECD = Organisation of Economic Co-operation and Development; T&D = transmission and distribution; VIU = vertically integrated utility WDI = World Development Indicators.
a. ESMAP META Model.
b. Eberhard and others 2011.

We now provide the methodology used to estimate collection rates in Oman, Saudi Arabia, and Qatar, for which we did not have direct data from the MED. The bill collection rate is defined as the income effectively collected during the year in relation to the income billed, and is calculated using equation C.1.

$$\text{Bill collection rate} = \frac{\text{Income effectively collected from customers for energy consumption and related service}}{\text{revenues related to energy consumption and service}}$$

(C.1)

When the collection rate was not available, it was calculated from the annual reports and financial statements of the utilities. In other words, the rate is the revenues collected divided by the billed amount. Since the annual reports do not provide a value for billed amounts, it was approximated as follows:

1. The income effectively collected is considered to be the figure of annual sales of, or annual revenues from, electricity in the financial statement.
2. The income not collected is considered as the receivables from customers, as stated in the financial report.

3. The billed amount is therefore the sum of what was not collected (the receivables) and what was actually collected (the sales revenue reflected in the financial report).
4. The collection rate is therefore calculated using equation C.2

$$\text{The collection rate} = \frac{\text{sales revenue}}{\text{sales revenue} + \text{receivables from customers}} \quad (C.2)$$

5. If the economy has several utilities, steps 1–4 above were applied to each utility and the average of all utilities was taken to be the economy collection rate.

This methodology was used to calculate the economy QFDs for Oman, Saudi Arabia, and Qatar, as shown in panels a, b, and c of table C.4.

Table C.4 Data and Sources Used for Calculating Collection Rates

Economy	Oman	Oman	Oman
a. Oman			
Utility name	Muscat Electricity Distribution Company	Majan Electricity Company	Mazoon Electricity Distribution Company
Source of data	Annual report 2013	Annual report 2013	Annual report 2013
Amounts due from private customers (RO)	33,562,000	17,357,000	20,344,000
Amounts due from government customers (RO)	13,610,000	6,029,000	5,776,000
Electricity sales to private customers (RO)	98,814,000	79,265,000	67,567,000
Electricity sales to government customers (RO)	37,479,000	10,221,000	18,815,000
Collection rate (%)	74	79	77
b. Saudi Arabia			
Economy	Saudi Arabia		
Utility name	Saudi Electricity Company (SEC)		
Source of data	SEC publication: electric data 2000–14		
Receivables from customers and revenues accrued net	Saudi riyal (SRI) 18,452,000,000		
Total electricity sales	SRI 32,878,000,000		
Collection rate (%)	64		
c. Qatar			
Economy	Qatar		
Utility name	KAHRAMAA		
Source of data	KAHRAMAA Annual Report 2013		
Accounts receivable	Qatari riyal (QR) 585,434,000		
Revenues from sale of electricity	QR 1,553,741,000		
Collection rate (%)	73		

Source: World Bank calculations; RO: Omani Riyal.

Methodology for Estimating the Economy-Level Cost Recovery Tariff and Collection Rates in Selected Economies

Economy-Level QFD

Cost-recovery tariffs were calculated using the basis of the economy fuel mix, and the levelized cost of electricity (LCOE) from different energy sources, as follows:

$$T_c = \text{Weighted } LCOE = (LCOE_{Coal} \times \%_{Coal}) + (LCOE_{Hydro} \times \%_{Hydro}) + \\ (LCOE_{N.gas} \times \%_{N.gas}) + (LCOE_{Fuel} \times \%_{Fuel}) + (LCOE_{Renewables} \times \%_{Renewables})$$

The shares of energy mix in each country used to compute the cost-recovery tariff are in table C.5. The LCOE values corresponding to each generation source are presented in table C.6. The Energy Sector Management Assistance Program's (ESMAP's) Model for Electricity Technology Assessment (META) considered 2010 as the base year; transmission and distribution (T&D) costs were not included and neither were environmental costs. To account for T&D losses, a value of US¢ 3.2 per kilowatt-hour (kWh) was added.

Table C.5 Share of Energy Mixes Used in the Calculation of Tc (%)

Economy	Coal	Hydro	Natural gas	Fuel	Renewables
Algeria	0	1	93	7	0
Bahrain	0	0	100	0	0
Djibouti	0	0	0	100	0
Egypt, Arab Rep.	0	8	77	15	1
Iraq	0	8	55	19	0
Jordan	0	0.3	25	74	0.1
Lebanon	0	7	0	93	0
Morocco	43	10	21	21	5
Oman	0	0	97	3	0
Qatar	0	0	100	0	0
Saudi Arabia	0	0	53	24	0
Tunisia	0	0.3	96	0.4	2
Yemen, Rep.	0	0	32	68	0
Israel[a]	54	0	42	36	1

Source: WDI.

a. in the case of West Bank, all electricity is imported from Israel, therefore the LCOE of Israel is used for T_c.

Table C.6 LCOE Values Used to Calculate the Cost-Recovery Tariffs and Their Sources

Generation type	LCOE (US$ cents) /kWh	Source
Coal	7.44	ESMAP META Model
Hydro	2.86	ESMAP META Model
Natural gas	8.12	ESMAP META Model
Fuel	31.45	Average Lazard
Renewables	6.9	Average Lazard[a]

Source: World Bank calculations based on ESMAP META Model and Lazard. 2014.
Note: ESMAP = Energy Sector Management Assistance Program; LCOE = levelized cost of electricity; META = Model for Electricity Technology Assessment.
a. Considering utility-sized photovoltaics (PV) and wind only.

Utility-Level QFD: Calculating the Unit Historical Cost

The unit historical cost, Tc, is composed of three main components that account for both capital expenditure (CAPEX) and operating expenses (OPEX). Equation C.3 below is used for calculating the unit historical cost made up of three annualized components:

$$Unit\ historical\ cost = \text{Infrastructure CAPEX + Connection CAPEX + OPEX} \tag{C.3}$$

Where:

$$Infrastructure\ \text{CAPEX} = \frac{cost\ for\ power\ generation + cost\ for\ T\&D}{0.95 \times (\text{kWh generated} + \text{kWh purchased})}$$

$$Connection\ \text{CAPEX} = \frac{\text{CAPEX of T\&D} \textit{ connection per customer} \times \textit{number of new customers}}{0.95 \times (\text{kWh generated} + \text{kWh purchased})}$$

$$\text{OPEX} = \frac{\text{Total OPEX from statements}}{\text{kWh billed}}$$

Differences between the economy-level and the utility-level QFD values can be explained primarily in the calculation of the cost-recovery tariff, Tc. While the economy-level QFD does not take into consideration the energy purchased and imported, and only considers the energy generated within an economy, this is accounted for in the utility-level QFD and can be observed in the cases of Djibouti[1] and the Republic of Yemen, for example. The effective tariff was approximated to that used previously in the economy-level QFD calculations, that is, for an average monthly consumption of 250 kWh in the residential sector.

The Infrastructure CAPEX component of the Unit Historical Cost

Using the CAPEX figures mentioned in the financial statements of utilities can be misleading in MENA. Since most utilities are public vertically integrated utilities (VIUs), the CAPEX is often obtained in the form of subsidies from the state and this is not always properly reflected in the financial statements.

The annualized CAPEX related to infrastructure is made up of the costs related to power generation and those related to investments in T&D infrastructure.

Calculating the Generation CAPEX

The annualized CAPEX for generation was calculated depending upon the installed capacity of the plant, the technology type, and its economic life. This is shown in equation C.4.

$$Amortized\ capital\ cost = Installed\ capacity \times \text{CAPEX}\ per\ kWh \times \frac{r}{1 - \dfrac{1}{(1+r)^{T}}} \tag{C.4}$$

Table C.7 Components Used to Calculate the CAPEX According to Technology Type

Technology	CAPEX[a] per kW ($/kW)	Economic life (T)	Amortization factor $r \div [\, 1 - (1 + r)^{-T}\,]$
Biomass	2,500	30	0.1062
Coal	2,403	30	0.1062
Co-Gen	917	30	0.1062
Diesel	1,070	30	0.1062
Gas CC	917	30	0.1062
Gas OC	603	30	0.1062
Geothermal	4,362	30	0.1062
HFO	1,250	30	0.1062
Hydropower	1,500	35	0.1037
Nuclear	4,102	60	0.1003
Solar	2,500	25	0.1102
Wind	2,000	25	0.1102

Source: Trimble and others 2016.
Note: CAPEX = capital expenditure; CC = combined cycle; HFO = heavy fuel oil; kW = kilowatts; OC = open cycle.
a. World Bank 2016b: 70.

Data were used from table C.7, as well as from data on the installed capacities of the utilities involved. A discount rate (r) of 10 percent was used and the assumption was made that the hydro plants in MENA are big since the MENA region does not have a significant amount of hydro installed capacity. This avoided a degree of complexity in this calculation.

Calculating the T&D CAPEX

The annualized CAPEX for the T&D infrastructure depends upon the type of line voltage, its economic life, and the length of the network as shown in table C.8. This was calculated using a discount rate (r) of 12 percent and equation C.5:

$$Amortized\ capital\ cost = Length\ of\ T\&D\ network \times CAPEX\ per\ km \times \frac{r}{1 - \dfrac{1}{\left(1+r\right)^{T}}} \qquad (C.5)$$

Table C.8 Components Used to Calculate the CAPEX of the T&D Network

	Assumed CAPEX ($/km)	Economic life in years (T)	Amortization factor $r \div [\, 1 - (1 + r)^{-T}\,]$
Lines 110 kV or above	165,000	50	0.1204
Lines below 110 kV down to 66 kV	65,000	40	0.1213
Lines below 66 kV down to 1 kV	10,000	30	0.1241

Source: World Bank 2016b: 71.
Note: CAPEX = capital expenditure; km = kilometers; kV = kilovolts; T&D = transmission and distribution.

The Customer Connection CAPEX component of the Unit Historical Cost

The cost of connecting a customer is calculated using equation C.6:

$$\frac{\text{CAPEX of T\&D } \textit{connection per customer } \times}{\textit{Total number of new customers}} \over 0.95 \times (\text{kWh generated} + \text{kWh purchased})} \qquad (C.6)$$

The CAPEX related to T&D for connecting each customer is considered to be $100. Multiplying the electricity generated plus the energy the utility purchases, 0.95 corresponds to the ratio of electricity actually dispatched if the normative losses are considered to be 5 percent.

The OPEX component of the Unit Historical Cost

Annualized OPEX is expressed as a share of the electricity generated, as shown in equation C.7 below:

$$\frac{\text{Total OPEX from statememts}}{\textit{Electricity billed}} \qquad (C.7)$$

Estimating Labor Costs for the Arab Republic of Egypt, Djibouti, Jordan, Morocco, Oman, and the Republic of Yemen

Labor costs were unavailable for several utilities and were obtained based on calculations making use of an estimated unit labor cost, as described for the countries listed below.

Egypt

Egypt has an unbundled electricity sector with a total of 12 utilities (including generation, distribution, and transmission) under the Egyptian Electricity Holding Company (EEHC). The cost of labor for all utilities except the Hydro Power Plants Electricity Production Company was available in the MENA database. However, the number of employees for all the utilities was not available.

To calculate the total cost of labor, including that of the Hydro Power Plants Electricity Production Company and EEHC, a unit average cost per employee was calculated from the data for the utilities with labor costs and number of employees available. This unit cost was then multiplied by the total number of employees to obtain the value for the total labor cost for Egypt. The values used are found in table C.9.

Table C.9 Values and Methodology Used in Calculating Labor Costs for the Arab Republic of Egypt

Equation	Description	Value
A	Number of employees without the Hydro Power Plants Electricity Production Company and without EEHC	172,733
B	Cost of labor in all utilities except the EEHC and Hydro Power Plants Electricity Production Company	$1,359,678,577
$C = B/A$	Unit cost of labor	$7,872
D	Number of employees in the EEHC	3,586
E	Number of employees in the Hydro Power Plants Electricity Production Company	3,038
$F = (D+E) \times C$	Cost of employees in the EEHC and Hydro Power Plants Electricity Production Company	$52,141,228
$G = F+B$	Total estimated cost of labor including EEHC and Hydro Power Plants Electricity Production Company	$1,411,819,806

Source: MENA Electricity Database and World Bank calculations.
Note: EEHC = Egyptian Electricity Holding Company; MENA = Middle East and North Africa.

Jordan

For Jordan, several utilities had no data. Initially out of 10 utilities, the number of employees was available for 9, and labor costs for 8. Data from the report of the Jordanian regulator, the Energy and Minerals Regulatory Commission (EMRC), were used for the utility with the missing number of employees (Qatrana Electric Power Company, QEPC). Since the Amman Asia utility was not operational in the year of study (2013), it was neglected.

As a result, there were nine utilities with nine employee numbers available, and seven with labor costs available. The same methodology as used in the case of Egypt was applied here to calculate the total cost of labor for the nine utilities in Jordan. The number of employees and labor costs values available are shown in table C.10.

Table C.11 shows the total number of employees and the total labor costs for the 7 utilities for which data was available in Jordan.

Table C.10 Utilities and Data Available for Jordanian Utilities

		No. employees	Labor costs in $
1	AES Levant Holding B.V.	47	Not available
2	Amman East Power Plant (AES)	51	3,248,314
3	Central Electricity Generating Company	1,037	18,788,759
4	Electricity Distribution Company	1,320	19,813,536
5	Irbid District Electricity Company	1,088	16,270,190
6	Jordan Electric Power Company	2,602	86,150,700
7	National Electric Power Company	1,373	22,166,850
8	Qatrana Electric Power Company	78	Not available
9	Samra Electric Power Generation Company	345	6,096,730

Source: MENA Electricity Database.

Table C.11 Calculating the Unit Labor Cost for Jordan

Description	Value
Total number of employees in 7 utilities	7,941
Total labor costs for 7 utilities	$175,535,079
Unit labor cost calculated using equation C.6	$22,075

Source: World Bank calculations based on MENA Electricity Database.

Note that total labor costs and total number of employees in equation C.8 do not include data for AES Levant Holding B.V and Qatrana Electric Power Generation Company.

$$\text{Unit labor cost} = \frac{\text{Total labor costs}}{\text{Total number of employees}} \qquad (C.8)$$

Equation C.9 is used to calculate the labor costs of the two utilities with missing values and table C.12 shows the results obtained.

$$\text{Estimated labor cost} = \text{Number of employees in Utility} \times \text{Unit labor cost} \qquad (C.9)$$

The total cost of labor for Jordan is shown in table C.13 and was obtained using the equation C.10.

$$\text{Total labor cost} = \text{Total labor costs for 7 utilities with data} + \text{labor costs of remaining two utilities} \qquad (C.10)$$

Table C.12 Calculating the Cost of Labor for the Two Utilities with Missing Values for Jordan

Description	Value ($)
Estimated labor costs for Qatrana	1,721,819
Estimated labor costs for AES Levant	1,037,506

Source: World Bank calculations based on MENA Electricity Database.

Table C.13 Calculating the Total Labor Costs for Jordan

Description	Value ($)
Total cost of labor for 9 utilities	175,294,404

Source: World Bank calculations based on MENA Electricity Database.

Morocco

In the case of Morocco, the number of employees of all the utilities were available, but not the labor costs per utility. This is shown in table C.14 while table C.15 details how the unit labor cost was obtained.

Table C.16 shows how the missing values were obtained using values from table C.14.

Finally, the total cost of labor for all the utilities for Morocco is shown in table C.17.

Table C.14 Utilities and Data Available for Moroccan Utilities

		No. employees (A)	Labor costs in $ (B)
1	AMENDIS Tanger	401	25,306,122
2	AMENDIS Tetouan	468	25,772,595
3	LYDEC	1,432	92,912,657
4	ONEE	8,796	252,453,751
5	RADEEL	134	
6	REDAL	511	44,702,600
7	Regie de Kenitra	196	
8	Regie de Marrakech	370	8,355,024
9	Regie de Meknes	208	
10	RADEEJ	188	4,131,731
11	Regie de Fes	439	
12	Regie de Safi	118	

Source: MENA Electricity Database.

Table C.15 Calculating the Unit Labor Cost for Morocco

Equation	Description	Value
$C = \sum\limits_{A1}^{A12}$	Total number of employees available	13,261
$D = \sum\limits_{B1}^{B12}$	Total labor costs available	$453,634,480
$E = D/C$	Unit labor cost	$34,208

Source: World Bank calculations based on MENA Electricity Database.

Table C.16 Calculating the Cost of Labor for the Utilities with Missing Values for Morocco

Equation	Description	Value
$F = E \times (A5 + A9 + A11 + A12)$	Labor cost in remaining utilities	$37,457,940

Source: World Bank calculations based on MENA Electricity Database.

Table C.17 Calculating the Total Labor Costs for Morocco

Equation	Description	Value
$G = D + F$	Total cost of labor for all utilities	$491,092,421

Source: World Bank calculations based on MENA Electricity Database.

Oman

Omani utilities often have a larger number of outsourced employees than full-time employees. For consistency, it was decided to include the total number of employees in the labor cost estimates. Twelve utilities were found to have data for both the total number of employees and the labor costs. A unit cost of labor was calculated from these twelve utilities, which was equal to $19,742.

The total number of employees for the 12 utilities was 5,085 and the total cost of labor obtained for these 12 utilities was $100,385,560. After obtaining an aggregate value of the total direct and indirect employees in 2013 from the Authority for Electricity Regulation (AER) annual report for 2014, it can be estimated that the remaining number of employees (8,277 − 5,085), is 3,192.

The total estimated labor costs in Oman = (3,192 x 19,74) + $100,385,560
= $163 million

Republic of Yemen

The cost of labor for the Public Electricity Corporation (PEC), the Yemeni public VIU, was unavailable. Data on the number of employees were obtained. An estimate of the cost of labor was done using average values from the International Labor Organization (ILO) for the Republic of Yemen. This is shown in table C.18.

Using an exchange rate of $1 = 203.4 Yemeni riyals (corresponding to January 1, 2013), the values listed in table C.19 were obtained for the average unit annual cost of labor.

Table C.18 Calculating Average Monthly Earning Based upon ILO Data for the Republic of Yemen

Position	Monthly salary in YRls (Yemeni Riyals)
Managers	30,290
Clerical support workers	42,591
Technicians and associate professionals	69,439
Average monthly earning calculated	47,440

Source: World Bank calculations based on International Labor Organization.
Note: ILO = International Labor Organization.

Table C.19 Calculating the Cost of Labor for the Republic of Yemen

Assuming salary paid for 12 months	
Average annual cost in U.S. dollars per employee	$2,797
Number of employees in PEC	18,126
Total estimated salary bill in U.S. dollars (cost of labor)	**$50,706,483**

Source: World Bank calculations based on MENA Electricity Database and ILO.
Note: PEC = Public Electricity Corporation.

Note

1. In the case of Djibouti, the electricity volume billed was assimilated to the energy generated + energy imported, since the figures obtained otherwise did not seem to reflect the fact that Djibouti imported an amount equivalent to 73 percent of the electricity generated in the economy. A factor of 0.86 was added to account for the 16 percent system losses in Djibouti.

References

AER (Authority for Electricity Regulation). 2013. *Annual Report 2013*. http://www.aer-oman.org/pdfs/Annual%20Report%202013%20-%20Eng.pdf

Arab Union of Electricity. 2014. "Electricity Tariff in Arab Countries."

Eberhard, A., O. Rosnes, M. Shkaratan, H. Vennemo. 2011. Africa's Power Infrastructure: Investment, Integration, Efficiency. Directions in Development; infrastructure. Washington, DC: World Bank. https://openknowledge.worldbank.org/handle/0986/2290.

EUEI (European Union Energy Initiative). 2013. *Country Power Market Brief: Djibouti*. Africa-EU Energy Partnership http://www.euei-pdf.org/sites/default/files/field_publication_file/AEEP_Djibouti_Country_market_brief_EN.pdf.

EEHC (Egyptian Electricity Holding Company). 2014. *Annual Report 2013/2014*. http://www.moee.gov.eg/english_new/EEHC_Rep/REP-EN2013-2014.pdf.

ESMAP (Energy Sector Management Assistance Program). ESMAP Meta Model. http://esmap.org/META.

Iraq Energy Institute. 2015. "Iraq's Biggest Power Threat." Iraq Energy Forum, August 6, 2015, Iraq Energy Institute, Baghdad. http://iraqenergy.org/home/articles_details.php?id=5.

KAHRAMMA. Annual Report 2013.

———. Sustainability Report 2013.

———. Annual Report 2014.

Lazard. 2014. "Lazard's Levelized Costs of Energy Analysis, version 8.0." https://www.lazard.com/media/1777/levelized_cost_of_energy_-_version_80.pdf.

Lebanon Ministry of Environment and United Nations Development Programme (UNDP). http://climatechange.moe.gov.lb/viewfile.aspx?id=64.

NEPCO (National Electric Power Company). 2013. *Annual Report 2013*. Jordan: NEPCO.

SEC. "Statistics 2000 to 2014."

Trimble, C., M. Kojima, I. P. Arroyo, and F. Mohammadzadeh. 2016. "Financial Viability of Electricity Sectors in Sub-Saharan Africa: Quasi-Fiscal Deficits and Hidden Costs." Policy Research Working Paper 7788, World Bank, Washington, DC.

World Bank. 2013. World Development Indicators (database). World Bank, Washington, DC. https://data.worldbank.org/data-catalog/world-development-indicators.

———. 2014a. "YEM Power Ministerial Note." Unpublished paper, World Bank, Washington, DC. January 3.

———. 2014b. *West Bank and Gaza: Assessment and Action Plan to Improve Payment for Electricity Services in the Palestinian Territories: Study on Electricity Sector Contribution to Net Lending*. Report No: ACS9393. Washington, DC: World Bank.

http://documents.worldbank.org/curated/en/120271468317065014/pdf/ACS93930WP0P1469990Box385388B00OUO090.pdf.

———. 2016a. "eC2: Electricity Services Restoration and Operations Efficiency." ToR for World Bank Assignment Title: 1223732—IRAQ, Netherlands for the World Bank, July 24. https://nl4worldbank.org/2016/07/14/ec2electricity-services-restoration-and-operations-efficiency.

———. 2016b. "Financial Viability of Electricity Sectors in Sub-Saharan Africa: Quasi-Fiscal Deficits and Hidden Costs." Policy Research Working Paper WPS 7788, World Bank Group, Washington, DC.

Methodology for the Analysis of Drivers of Performance

There are several equivalent approaches to testing the equality of the means of two subgroups of observations. This study uses regression on dummy variables because it is immediately generalizable to testing for the equality of three (or more) subgroup means and to testing equality for means of each subgroup when two or more subgroups are analyzed at the same time.

The dummy variable approach defines a variable of interest (the benchmark indicator) whose observations can be ascribed to one of two subgroups using a 1/0 classification. For example, data were available for the load factor—denoted $YL(i)$ for the i'th observation—for 23 utilities, of which six were vertically integrated utilities (VIUs) and 17 were distribution utilities (DUs). The null hypothesis is that there is no structure effect so that the means of the two groups are equal. The mean load factor for the first group was 0.567, and for the second it was 0.554. Rather than use the standard test statistic for the equality of two means based on these values and the estimated variance for the pooled sample, a regression approach can be used.

Let $D1(i)$ take the value 1 if the observation is from a VIU and zero if it is from a DU, and let $D2(i)$ take the value 1 if it is from a distribution utility and value 0 if it is from a VIU. The regression model expresses the load factor in terms of the two dummy variables and an error term as shown in equation (D.1):

$$YL(i) = \beta_1 D1(i) + \beta_2 D2(i) + u(i) \tag{D.1}$$

where the values of the β are to be estimated. This equation can be interpreted as saying that all values of the load factor for VIUs are equal to $\beta_1 + u(i)$, while for DUs they are equal to $\beta_2 + u(i)$. Assuming that the means of the error term are zero for both groups, estimates of the β values can be obtained by ordinary least squares and are denoted b_1 and b_2. It is important to note that this equation does not include a constant. Indeed, attempting to estimate an equation with a constant and two dummy variables that correspond to the two states of the dichotomy under consideration would lead to exact singularity of the data matrix. The values obtained are $b_1 = 0.557$ and $b_2 = 0.554$, the same as obtained by simply finding means for the two sets of observations. The test for the equality of the two means is then equivalent to a test for the equality of the β in the regression model. A test of a linear restriction ($\beta_1 = \beta_2$) for a regression model is provided by Wald's test based on an F statistic. For the load factor, the Wald test indicates that the probability of observing a difference between them at least as large as that estimated is 0.80. That is, for the load factor, there is an 80 percent chance of observing a difference between the subgroup mean load factors of 0.003 or greater. A probability of 5 percent is regarded as indicating a significant result that supports the alternative hypothesis that the group means are not equal. A 10 percent probability is regarded as being worthy of note, if not highly significant.

The same results can be obtained through a different formulation of the dummy variable model. A dummy variable $DC(i)$ is defined as taking the value 1 for all observations, while $D2(i)$ is defined as before. The model is now written as in equation (D.2):

$$YL(i) = \gamma_1 \, DC(i) + \gamma_2 \, D2(i) + u(i) \qquad\qquad (D.2)$$

Noting that $DC(i)$ is constant for all observations, equation (D.2) is equivalent to equation (D.3):

$$YL(i) = \gamma_1 + \gamma_2 \, D2(i) + u(i) \qquad\qquad (D.3)$$

which corresponds to a single variable regression model with a constant.

The interpretation of this model is that observations for VIUs have a mean of γ_1, while those for distribution utilities (DU) have a mean of $\gamma_1 + \gamma_2$. The coefficient on the DU (γ_2) is now the difference from the VIU. A standard t test for the hypothesis that this difference is zero is equivalent to a test of the equality of the two means. In the example above, $g_1 = 0.567$ and $g_2 = -0.013$, and the probability level for the t statistic on the difference coefficient g_2 is 0.80.

Certain factors in the study were categorized as falling into one of three classes (for example, big, medium, or small). The null hypothesis that the mean load factor is the same for all three groups is tested by constructing three dummy variables ($D1 = 1$ for big, $= 0$ for medium or small; $D2 = 1$ for medium, $= 0$ for big or small; $D3 = 1$ for small, $= 0$ for big or medium). From the regression of the load factor on these three variables the coefficients are equal to the subgroup means. Equality of the three means can be tested with a Wald test based on two linear restrictions ($\beta_1 = \beta_2$; $\beta_2 = \beta_3$).

For testing the effect of more than one categorization on the indicator of interest, it is simplest to use the approach of equation (D.2). Consider the case of the load factor in which both structure and ownership are to be considered at the same time. There are four different combinations of states: publicly owned VIU, privately owned VIU, publicly owned DU, and privately owned DU. The model includes a constant factor, a dummy variable for the DU, and a dummy variable for privately owned DU. The three variables fully define all four states. A publicly owned VIU takes the value of the constant, while DUs (whether public or private) have an incremental effect given by the coefficient on the distribution dummy, and private utilities (whether VIU or DU) have an incremental effect given by the coefficient on the ownership dummy. The hypothesis test that both the structure effect and ownership effect are zero can be carried out by a Wald test ($\beta_1 = \beta_2$; $\gamma_1 = \gamma_2$). This approach can be easily generalized to the case where all five factors are included and where some factors (size, income) are categorized into three states.

For indicators in which there are three states (size, income), the Wald test is carried out in two stages. First, the two restrictions $\beta_1 = \beta_2$ and $\beta_2 = \beta_3$ are simultaneously tested, and then the pairwise restrictions $\beta_1 = \beta_2$, $\beta_1 = \beta_3$, and $\beta_2 = \beta_3$ are tested one at a time, so as to identify which variables (if any) have different means from the others.

Core Values for MENA Indicators

This appendix provides the values used for the static analysis of this report, that is, year 2013 or where data were missing for that year, the most recent year for which data were available between 2009 and 2012. Table E.1 lists the abbreviations and full names of indicators used in the following tables E.2, E.3, and E.4. These tables report values of the latest year available for MENA electricity utilities, for technical and operational indicators, financial indicators, and commercial indicators, respectively.

Table E.1 Indicator Names and Their Abbreviations, as Used in Tables E.2–E.4 of This Appendix

	Indicator name	*Abbreviated name*
Technical and operational indicators	Load factor	Load Factor
	Capacity factor	Cap. Factor
	Availability factor	Av. Factor ·
	Transmission losses	Tran. Losses
	Distribution losses	Dis. Losses
	Technical losses	Tech. Losses
	Nontechnical losses	N.tech Losses
	Network maintenance	Network maint.
	# of meters replaced/total # of meters	Share of meters replaced
	Total OPEX/full-time equivalent (FTE) employee	OPEX per emp.
	Total OPEX per connection	OPEX per con.
	Total OPEX/kWh sold	OPEX per kWh sold
	Total OPEX/km of network	OPEX per km
	# of residential connections/FTE employee	# of res. con. per emp.
	Energy sales ($)/FTE employee	Energy sales per emp.
	Total revenues ($)/FTE employee	Total rev. per emp.
Financial indicators	Share of cost of (fuel, lubricant, gas and coal) in total OPEX	Share of cost of fuel in OPEX
	Share of (energy purchases and cost of fuel, lubricant, gas and coal) in total OPEX	Share of energy purchased in OPEX
	Share of labor cost in total OPEX	Labor cost in OPEX
	Energy sales/total OPEX	Engy sales/OPEX
	Energy sales/total costs	Engy sales/tot. costs
	(Accounts receivable/sales) × 365	Acc. rec./sales
	Debt/equity	Debt/equity
	Current assets/current liabilities	Assets/liab.
	ROA (return on assets)	ROA
	ROE (return on equity)	ROE
Commercial indicators	Total energy volume sold (kWh)/connection	Engy vol. sold per con.
	Residential energy volume sold (kWh)/connection	Res. engy vol. sold per con.
	Total billing ($)/connection	Billing per con.
	Residential billing ($)/connection	Res. billing per con.
	Collection rate	Collection rate
	Share of installed meters (%)	Share of installed meters
	SAIFI	SAIFI
	SAIDI	SAIDI
	CAIDI	CAIDI
	Duration of interruption taken into consideration for system interruptions affecting customers (including SAIDI, SAIFI, and CAIDI customer measures).	Duration of interruption

Source: World Bank calculations.
Note: CAIDI = Customer Average Interruption Duration Index; km = kilometer; kWh = kilowatt-hours; OPEX = operating expenses; SAIDI = System Average Interruption Duration Index; SAIFI = System Average Interruption Frequency Index.

Table E.2a Technical and Operational Indicators

| | | | Technical and operational indicators | | | | | | | | |
| | | | System and operational efficiency | | | | | | | | |
Country or economy	Utility type	Utility	Load factor (%)	Cap. factor (%)	Av. factor (%)	Tran. losses (%)	Dis. losses (%)	Tech. losses (%)	N.tech losses (%)	Network maint. (%)	Share of meters replaced (%)
Algeria	VIU	SONELGAZ	50	29			19	10	11		
Bahrain	VIU	EWA	52		96					6	
Djibouti	VIU	EDD								0.4	
Egypt, Arab Rep.	DU	AEDC	61	n.a.	n.a.	n.a.	11	7	4	n.a.	n.a.
	GU	CEPC	n.a.	58	n.a.	n.a.	n.a.	n.a.	n.a.	n.a.	n.a.
	DU	CEDC	38	n.a.	n.a.	n.a.	6	4	3	n.a.	n.a.
	GU	EDEPC	n.a.	60		n.a.	n.a.	n.a.	n.a.	n.a.	n.a.
	TU	EETC	n.a.	n.a.	n.a.			n.a.	5		n.a.
	DU	EEDC	62	n.a.	n.a.		10	5	5		
	GU	MDEPC	n.a.	65		n.a.	n.a.	n.a.	n.a.	n.a.	n.a.
	DU	MEEDC	69	n.a.	n.a.	n.a.	11	4	4		
	DU	NCEDC	62	n.a.	n.a.	n.a.	10	5	4		
	DU	NDEDC		n.a.	n.a.	n.a.	9	7	4		
	DU	SCEDC	64	n.a.	n.a.	n.a.	8	6	2		
	DU	SDEDC	60	n.a.	n.a.	n.a.	10				
	DU	UEEDC	68	n.a.	n.a.	n.a.	8				
	GU	UEEPC	n.a.	70		n.a.	n.a.	n.a.	n.a.		
	GU	WDEPC	n.a.	57		n.a.	n.a.	n.a.	n.a.	n.a.	n.a.
Iraq	VIU	MoE					37			n.a.	n.a.
Jordan	GU	AES Levant	n.a.	28	99	n.a.	n.a.	n.a.	n.a.	n.a.	n.a.
	GU	AAEPC	n.a.			n.a.	n.a.	n.a.	n.a.	n.a.	n.a.
	GU	AES PSC	n.a.	80	97	n.a.	n.a.	n.a.	n.a.	n.a.	n.a.
	GU	CEGCO	n.a.	50		n.a.	n.a.	n.a.	n.a.	n.a.	n.a.
	DU	EDCO	n.a.	n.a.	n.a.	n.a.	12	n.a.	n.a.	n.a.	n.a.

table continues next page

Table E.2a Technical and Operational Indicators *(continued)*

Country or economy	Utility type	Utility	Load factor (%)	Cap. factor (%)	Av. factor (%)	Tran. losses (%)	Dis. losses (%)	Tech. losses (%)	N.tech losses (%)	Network maint. (%)	Share of meters replaced (%)
	DU	IDECO		n.a.	n.a.	n.a.	11				1
	DU	JEPCO	51	n.a.	n.a.	n.a.	14	11	3	0	
	TU	NEPCO	n.a.	n.a.	n.a.	n.a.	n.a.	n.a.	n.a.	n.a.	n.a.
	GU	QEPCO	n.a.	75	97		n.a.	n.a.	n.a.	n.a.	n.a.
	GU	SEPCO	n.a.	50	n.a.	n.a.	n.a.	n.a.	n.a.	n.a.	n.a.
Lebanon	VIU	EdL					33	15	17		
Morocco	DU	AMENDIS Tanger		n.a.	n.a.	n.a.	10	8	2		
	DU	AMENDIS Tetouan		n.a.	n.a.	n.a.	11	9	2		
	DU	LYDEC		n.a.	n.a.	n.a.	7				
	VIU	ONEE	66			4	15	8	6	1	1
	DU	RADEEL		n.a.	n.a.	n.a.	8				
	DU	REDAL	56	n.a.	n.a.	n.a.	8	5	3	1	
	DU	RAK		n.a.	n.a.	n.a.	8				
	DU	RADEEMA	51	n.a.	n.a.	n.a.	5			3	
	DU	RADEM		n.a.	n.a.	n.a.	7				
	DU	RADEEJ	64	n.a.	n.a.	n.a.	4				2.7
	DU	RADEEF		n.a.	n.a.	n.a.					
	DU	RADEES		n.a.	n.a.	n.a.	3				
Oman	GU	APBS	n.a.	59	93	n.a.	n.a.	n.a.	n.a.	n.a.	n.a.
	GU	ABPC	n.a.	41	96	n.a.	n.a.	n.a.	n.a.	n.a.	n.a.
	GU	ASPC	n.a.	32	90	n.a.	n.a.	n.a.	n.a.	n.a.	n.a.
	GU	GPDCO	n.a.	58	85	n.a.	n.a.	n.a.	n.a.	n.a.	n.a.

table continues next page

Table E.2a Technical and Operational Indicators *(continued)*

Country or economy	Utility type	Utility	Technical and operational indicators — System and operational efficiency								
			Load factor (%)	Cap. factor (%)	Av. factor (%)	Tran. losses (%)	Dis. losses (%)	Tech. losses (%)	N.tech losses (%)	Network maint. (%)	Share of meters replaced (%)
	GU	AKPP	n.a.	67	89	n.a.	n.a.	n.a.	n.a.	n.a.	n.a.
	GU	ARPP	n.a.			n.a.	n.a.	n.a.	n.a.	n.a.	n.a.
	GU	BPDP	n.a.			n.a.	n.a.	n.a.	n.a.	n.a.	n.a.
	VIU	DPC					15			0	1.5
	DU	MJEC	71	n.a.	n.a.	n.a.	13	7	6		
	DU	MZEC	44	n.a.	n.a.	n.a.	11				
	DU	MEDC	55	n.a.	n.a.	n.a.	9	5	5		
	TU	OETC	n.a.	n.a.	n.a.	3	n.a.	n.a.	n.a.	n.a.	n.a.
	GU	PPC	n.a.			n.a.	n.a.	n.a.	n.a.	n.a.	n.a.
	VIU	RAECO					11			1	2.1
	GU	SSPWC	n.a.			n.a.	n.a.	n.a.	n.a.	n.a.	n.a.
	GU	SPP	n.a.	69	93	n.a.	n.a.	n.a.	n.a.	n.a.	n.a.
	GU	UPC	n.a.	50	91	n.a.	n.a.	n.a.	n.a.	n.a.	n.a.
	GU	WAJPCO	n.a.			n.a.	n.a.	n.a.	n.a.	n.a.	n.a.
Qatar	VIU	KAHRAMAA									
Saudi Arabia	VIU	SEC	57				5			0.1	
Tunisia	VIU	STEG	61	37		2	14	8	7		0.4
West Bank	DU	JDECO		n.a.	n.a.	n.a.	26				
	DU	NEDCO	38	n.a.	n.a.	n.a.	13				9.1
	DU	TUBAS	28	n.a.	n.a.	n.a.	16	11	5	3	1
Yemen, Rep.	VIU	PEC	55	46			36				0

Note: DU = distribution utility; GU = generation utility; n.a. = not applicable; TU = transmission utility; VIU = vertically integrated utility.

223

Table E.2b Technical and Operational Indicators

Country or economy	Utility type	Utility	Technical and operational indicators						
			Cost-efficiency (Total OPEX)				Labor efficiency		
			OPEX per emp.	OPEX per con.	OPEX per kWh sold	OPEX per km	# of res. con. per emp.	Energy sales per emp.	Total rev. per emp.
			$/emp.	$/con.	$/kWh	$/km	#con./emp.	$/emp.	$/emp.
Algeria	VIU	SONELGAZ	31,050	304	0.05	7,730			
Bahrain	VIU	EWA			0.08				
Djibouti	VIU	EDD	79,469	1,612	0.43	75,869	44	121,124	121,124
Egypt, Arab Rep.	DU	AEDC	23,911	134	0.04	15,207	155	19,920	22,783
	GU	CEPC	138,270	n.a.	n.a.	n.a.	n.a.	n.a.	n.a.
	DU	CEDC	47,446	230	0.04	10,406	178	42,083	43,943
	GU	EDEPC	91,535	n.a.	n.a.	n.a.	n.a.	n.a.	n.a.
	TU	EETC		n.a.	n.a.	n.a.	n.a.	n.a.	n.a.
	DU	EEDC	34,612	157	0.04	8,513	188	31,038	34,683
	GU	MDEPC	75,610	n.a.	n.a.	n.a.	n.a.	n.a.	n.a.
	DU	MEEDC	37,360	115	0.03	6,199	296	32,182	37,258
	DU	NCEDC	46,455	157	0.04	12,780	252	42,052	44,462
	DU	NDEDC	38,554	101	0.03	8,502	315	38,420	40,756
	DU	SCEDC	46,892	169	0.04	14,788	233	43,406	45,113
	DU	SDEDC	27,665	75	0.03	9,923	319	28,216	29,709
	DU	UEEDC	37,974	119	0.03	6,373	287	33,131	38,279
	GU	UEEPC	178,678	n.a.	n.a.	n.a.	n.a.	n.a.	n.a.
	GU	WDEPC	77,375	n.a.	n.a.	n.a.	n.a.	n.a.	n.a.
Iraq	VIU	MOE		820	0.07	30,885	n.a.	n.a.	n.a.
Jordan	GU	AES Levant		n.a.	n.a.	n.a.	n.a.	n.a.	n.a.
	GU	AAEPC	311,051	n.a.	n.a.	n.a.	n.a.	n.a.	n.a.
	GU	AES PSC		n.a.	n.a.	n.a.	n.a.	n.a.	n.a.
	GU	CEGCO		n.a.	n.a.	n.a.	n.a.	n.a.	n.a.
	DU	EDCO	230,005			26,497	126	223,713	228,204
	DU	IDECO	196,963	547	0.10	11,900	310	211,193	

table continues next page

Table E.2b Technical and Operational Indicators (continued)

Country or economy	Utility type	Utility	Technical and operational indicators						
			Cost-efficiency (Total OPEX)				# of res. con. per emp.	Labor efficiency	
			OPEX per emp.	OPEX per con.	OPEX per kWh sold	OPEX per km		Energy sales per emp.	Total rev. per emp.
			$/emp.	$/con.	$/kWh	$/km	#con./emp.	$/emp.	$/emp.
	DU	JEPCO	447,628		0.14	43,084	364	416,395	423,073
	TU	NEPCO		n.a.	n.a.	n.a.	n.a.	n.a.	n.a.
	GU	QEPCO		n.a.	n.a.	n.a.	n.a.	n.a.	n.a.
	GU	SEPCO		n.a.	n.a.	n.a.	n.a.	n.a.	n.a.
Lebanon	VIU	EdL	573,990	1,575	0.29			155,298	162,865
Morocco	DU	AMENDIS Tanger	321,620	508	0.12	35,835			355,728
	DU	AMENDIS Tetouan	151,086	346	0.15	30,610			142,100
	DU	LYDEC	527,093	836	0.20	96,152		527,759	565,777
	VIU	ONEE	283,710	510	0.09	9,362		333,564	347,685
	DU	RADEEL	202,559	361	0.12	18,849		174,709	221,208
	DU	REDAL	642,046	644	0.17	49,382	969	663,010	736,208
	DU	RAK	247,581	412	0.12	19,758		231,793	305,496
	DU	RADEEMA	287,299	410	0.10	31,932			372,131
	DU	RADEM	253,858	309	0.11	21,096		246,725	305,224
	DU	RADEEJ	190,092	396	0.10	21,285	403	258,237	258,791
	DU	RADEEF	185,909	318	0.11	37,114		182,504	230,975
	DU	RADEES	199,577	339	0.13	32,438		177,773	454,427
Oman	GU	APBS		n.a.	n.a.	n.a.	n.a.	n.a.	n.a.
	GU	ABPC	675,865	n.a.	n.a.	n.a.	n.a.	n.a.	n.a.
	GU	ASPC	816,078	n.a.	n.a.	n.a.	n.a.	n.a.	n.a.
	GU	GPDCO	304,725	n.a.	n.a.	n.a.	n.a.	n.a.	n.a.

table continues next page

Table E.2b Technical and Operational Indicators (continued)

			Technical and operational indicators						
			Cost-efficiency (Total OPEX)				Labor efficiency		
Country or economy	Utility type	Utility	OPEX per emp. $/emp.	OPEX per con. $/con.	OPEX per kWh sold $/kWh	OPEX per km $/km	# of res. con. per emp. #con./emp.	Energy sales per emp. $/emp.	Total rev. per emp. $/emp.
	GU	AKPP		n.a.	n.a.	n.a.	n.a.	n.a.	n.a.
	GU	ARPP		n.a.	n.a.	n.a.	n.a.	n.a.	n.a.
	GU	BPDP		n.a.	n.a.	n.a.	n.a.	n.a.	n.a.
	VIU	DPC	328,856	1,438	0.05	19,460	173	265,289	370,361
	DU	MJEC	226,672		0.05		92	157,010	273,953
	DU	MZEC	174,580	1,150		14,107	115	107,185	227,820
	DU	MEDC		1,698		42,246	399		
	TU	OETC	58,413	n.a.	n.a.	8,990	n.a.	n.a.	n.a.
	GU	PPC		n.a.	n.a.	n.a.	n.a.	n.a.	n.a.
	VIU	RAECO	89,332	4,917	0.21	30,847	13	23,179	107,349
	GU	SSPWC	614,812	n.a.	n.a.	n.a.	n.a.	n.a.	n.a.
	GU	SPP	942,235	n.a.	n.a.	n.a.	n.a.	n.a.	n.a.
	GU	UPC	288,524	n.a.	n.a.	n.a.	n.a.	n.a.	n.a.
	GU	WAJPCO		n.a.	n.a.	n.a.	n.a.	n.a.	n.a.
Qatar	VIU	KAHRAMAA		1,519					
Saudi Arabia	VIU	SEC	278,984	1,237	0.03	16,992	179	276,918	300,451
Tunisia	VIU	STEG	258,487	948		20,409		142,249	142,249
West Bank	DU	JDECO			0.19				
	DU	NEDCO	125,720	684	0.16	15,350	147	135,888	144,991
	DU	TUBAS		759	0.10	22,697	73	79,724	56,434
Yemen, Rep.	VIU	PEC	16,590	158	0.06	16,712	90		18,630

Source: MENA Electricity Database.

Note: DU = distribution utility; GU = generation utility; km = kilometer; kWh = kilowatt-hours; n.a. = not applicable; OPEX = operation expenses; TU = transmission utility; VIU = vertically integrated utility.

Table E.3 Financial Indicators

			Financial indicators									
			Cost structure			Cost-recovery		Balance sheet			Profitability	
Country or economy	Utility type	Utility	Share of cost of fuel in OPEX	Share of energy purchased in OPEX	Labor cost in OPEX	Engy sales/ OPEX	Engy sales/Tot. costs	Acc. rec./ sales	Debt/ equity	Assets/ liab.	ROA	ROE
			%	%	%	%	%	days	%	%	%	%
Algeria	VIU	SONELGAZ				92	56		428	146	-1.74	-7
Bahrain	VIU	EWA		67	5	37		205	67	84	0.88	1
Djibouti	VIU	EDD	22	56	17	152	110	192	222	274		
Egypt, Arab Rep.	DU	AEDC	n.a.	n.a.	8	83	n.a.	81		77	0.18	0.26
	GU	CEPC	88	n.a.		n.a.	n.a.	n.a.		5	0.0	1
	DU	CEDC	n.a.	n.a.	20	89	83	n.a.	685	66	2	8
	GU	EDEPC	88	n.a.	10	n.a.	n.a.	n.a.	3,484	37	0.05	0.3
	TU	EETC	n.a.	n.a.		n.a.	n.a.	n.a.		53		
	DU	EEDC	n.a.	n.a.	26	90	80	175	527	103	0.04	0.12
	GU	MDEPC	79	n.a.	12	n.a.	n.a.	n.a.	2,509	68	0.03	0.41
	DU	MEEDC	n.a.	n.a.	27	86	77	115	501	85	0.06	0.14
	DU	NCEDC	n.a.	n.a.	21	91	87	183	850	71	0.19	0.61
	DU	NDEDC	n.a.	n.a.	24		90		677	97	0.30	0.83
	DU	SCEDC	n.a.	n.a.	21	93	87	276	1,282	81	2.6	8.77
	DU	SDEDC	n.a.	n.a.	35				523	103	0.23	0.46
	DU	UEEDC	n.a.	n.a.	26	87	75	178	571	113	0.06	0.17
	GU	UEEPC	93	n.a.	5	n.a.	n.a.	n.a.	1,270	56	0.35	3.02
	GU	WDEPC	81	n.a.	15	n.a.	n.a.	n.a.	3,074	67	0.01	0.11
Iraq	VIU	MOE	30	84			n.a.	n.a.				
Jordan	GU	AES Levant		n.a.		n.a.	n.a.	n.a.				
	GU	AAEPC	63	n.a.		n.a.	n.a.	n.a.	290	123		
	GU	AES PSC		n.a.		n.a.	n.a.	n.a.	333	287		36
	GU	CEGCO		n.a.		n.a.	n.a.	n.a.	354	95	12	21

table continues next page

Table E.3 Financial Indicators (continued)

Country or economy	Utility type	Utility	Cost structure — Share of cost of fuel in OPEX %	Cost structure — Share of energy purchased in OPEX %	Cost structure — Labor cost in OPEX %	Cost-recovery — Engy sales/ OPEX %	Cost-recovery — Engy sales/Tot. costs %	Balance sheet — Acc. rec./ sales days	Balance sheet — Debt/ equity %	Balance sheet — Assets/ liab. %	Profitability — ROA %	Profitability — ROE %
	DU	EDCO	n.a.	n.a.	6	97		117	1,476	99	5	16
	DU	IDECO	n.a.	n.a.	7	107	99	120	981	84	6	20
	DU	JEPCO	n.a.	n.a.	7	93		122	576	80	25	12
	TU	NEPCO	n.a.	n.a.		n.a.	n.a.	n.a.	126			
	GU	QEPCO		n.a.		n.a.	n.a.	n.a.	621	488	5	25
	GU	SEPCO				n.a.	n.a.	n.a.	876	113	4	17
Lebanon	VIU	EdL	82	88	5	27	27			15	−150	
Morocco	DU	AMENDIS Tanger	n.a.	n.a.							3	3
	DU	AMENDIS Tetouan	n.a.	n.a.							−1	−2
	DU	LYDEC	n.a.	n.a.	12	100	89	76	279	72		18
	VIU	ONEE		81	10	118	87	159	3,327	63	−4	−127
	DU	RADEEL	n.a.	n.a.		86					6	7
	DU	REDAL	n.a.	n.a.	14	103	92	121		92	2	10
	DU	RAK	n.a.	n.a.		94						
	DU	RADEEMA	n.a.	n.a.	8	130		205	41			
	DU	RADEM	n.a.	n.a.		97						
	DU	RADEEJ	n.a.	n.a.	12	136	119	106	66	64	21	22
	DU	RADEEF	n.a.	n.a.		98						
	DU	RADEES	n.a.	n.a.		89						
Oman	GU	APBS	n.a.	n.a.		n.a.	n.a.	n.a.	249	121	14	16
	GU	ABPC	59	n.a.		n.a.	n.a.	n.a.	303	54	8	24
	GU	ASPC	61	n.a.		n.a.	n.a.	n.a.	294	53		
	GU	GPDCO	75	n.a.	13	n.a.	n.a.	n.a.	443	443	1	0.2

table continues next page

228

Table E.3 Financial Indicators *(continued)*

Country or economy	Utility type	Utility	Cost structure — Share of cost of fuel in OPEX %	Share of energy purchased in OPEX %	Labor cost in OPEX %	Cost-recovery — Engy sales/OPEX %	Engy sales/Tot. costs %	Acc. rec./sales days	Balance sheet — Debt/equity %	Assets/liab. %	Profitability — ROA %	ROE %
	GU	AKPP	78	n.a.		n.a.	n.a.	n.a.	94	79	9	15
	GU	ARPP	77	n.a.		n.a.	n.a.	n.a.		156		
	GU	BPDP	52	n.a.		n.a.	n.a.	n.a.	1,857	42	3	
	VIU	DPC			6	81	72	263	315	46		
	DU	MJEC	n.a.	n.a.	6	69		119	109	43	8	14
	DU	MZEC	n.a.	n.a.		61		110	148	18	6	14
	DU	MEDC	n.a.	n.a.	5	80		122	147	46	8	16
	TU	OETC	n.a.	n.a.		n.a.	n.a.	n.a.	192		7	20
	GU	PPC		n.a.		n.a.	n.a.	n.a.			0.1	
	VIU	RAECO	51	53	15			365	316	128	3	11
	GU	SSPWC	51	n.a.		n.a.		n.a.	357	179	3	13
	GU	SPP	68	n.a.		n.a.		n.a.	1,399	118	3	
	GU	UPC				n.a.		n.a.	72	38	5	7
	GU	WAJPCO	47	n.a.	25	n.a.	n.a.	n.a.		504	8	2
Qatar	VIU	KAHRAMAA	48		12	97		138	74	214		
Saudi Arabia	VIU	SEC			14	99	39	205	392	86	2	5
Tunisia	VIU	STEG	82	88	6	55	47	99	596	89	-4	-22
West Bank	DU	JDECO	n.a.	n.a.	10				260	126	-20	-19
	DU	NEDCO	n.a.	n.a.	7	108	103	166	86	275	3	4
	DU	TUBAS	n.a.	n.a.				276			7	
Yemen, Rep.	VIU	PEC	74	100						75		

Source: MENA Electricity Database.

Note: DU = distribution utility; GU = generation utility; n.a. = not applicable; OPEX = operation expenses; ROA = return on assets; ROE = return on equity; TU = transmission utility; VIU = vertically integrated utility.

Table E.4 Commercial Indicators

Country or economy	Utility type	Utility	Commercial indicators									
			Average consumption and billing					Metering	Customer management and service quality			
			Engy vol. sold per con.	Res. Engy vol. sold per con.	Billing per con.	Res. billing per con.	Collection rate	Share of installed meters	SAIFI	SAIDI	CAIDI	Duration of interruption
			kWh/con.	kWh/con.	$/con.	$/con.	%	%	000	minutes	minutes	minutes
Algeria	VIU	SONELGAZ	5,814									
Bahrain	VIU	EWA					97					
Djibouti	VIU	EDD	3,713	2,997			37					
Egypt, Arab Rep.	DU	AEDC	3,658	2,200	111	41	99	100	0.12	2.86	24.47	
	GU	CEPC	n.a.	n.a.	n.a.	n.a.	n.a.	n.a.	n.a.	n.a.	n.a.	n.a.
	DU	CEDC	5,862	2,194	197	41	94	100	0.17	4.59	27.12	
	GU	EDEPC	n.a.	n.a.	n.a.	n.a.	n.a.	n.a.	n.a.	n.a.	n.a.	n.a.
	TU	EETC	n.a.	n.a.	n.a.	n.a.	n.a.	n.a.				
	DU	EEDC	4,392	1,851	132	33	95	99	2.24	186.02	83.17	
	GU	MDEPC	n.a.	n.a.	n.a.	n.a.	n.a.	n.a.	n.a.	n.a.	n.a.	n.a.
	DU	MEEDC	3,746	2,686	96		92		0.50	58.30	117.76	
	DU	NCEDC	4,340	2,690	138	59	93		0.53	21.54	40.42	
	DU	NDEDC	3,133	2,323	97	50	84		0.93	17.26	18.48	
	DU	SCEDC	4,584	2,633	148	60	86	100	2.15		48.57	
	DU	SDEDC	2,438	1,839	68	33	93	100				
	DU	UEEDC	3,570	2,275	101	36	88	100	0.17	18.11	106.93	
	GU	UEEPC	n.a.	n.a.	n.a.	n.a.	n.a.	n.a.	n.a.	n.a.	n.a.	n.a.
	GU	WDEPC	n.a.	n.a.	n.a.	n.a.	n.a.	n.a.	n.a.	n.a.	n.a.	n.a.
Iraq	VIU	MOE	6,341		182		n.a.	n.a.	n.a.	n.a.	n.a.	
Jordan	GU	AES Levant	n.a.	n.a.	n.a.	n.a.	n.a.	n.a.	n.a.	n.a.	n.a.	n.a.
	GU	AAEPC	n.a.	n.a.	n.a.	n.a.	n.a.	n.a.	n.a.	n.a.	n.a.	n.a.
	GU	AES PSC	n.a.	n.a.	n.a.	n.a.	n.a.	n.a.	n.a.	n.a.	n.a.	n.a.
	GU	CEGCO	n.a.	n.a.	n.a.	n.a.	n.a.	n.a.	n.a.	n.a.	n.a.	n.a.
	DU	EDCO	6,429	3,487	335	335				12.71	n.a.	n.a.

table continues next page

Table E.4 Commercial Indicators (continued)

Country or economy	Utility type	Utility	Commercial indicators									
			Average consumption and billing					Metering	Customer management and service quality			
			Engy vol. sold per con.	Res. Engy vol. sold per con.	Billing per con.	Res. billing per con.	Collection rate	Share of installed meters	SAIFI	SAIDI	CAIDI	Duration of interruption
			kWh/con.	kWh/con.	$/con.	$/con.	%	%	000	minutes	minutes	
	DU	IDECO	5,591	3,054	586	247		100	3.84			
	DU	JEPCO	7,437	3,638		356	97		2.11	2.81		
	TU	NEPCO	n.a.	n.a.	n.a.	n.a.	n.a.	n.a.				
	GU	QEPCO	n.a.	n.a.	n.a.	n.a.	n.a.	n.a.	n.a.	n.a.	n.a.	n.a.
	GU	SEPCO	n.a.	n.a.	n.a.	n.a.	n.a.	n.a.	n.a.	n.a.	n.a.	n.a.
Lebanon	VIU	EdL	5,386		529							
Morocco	DU	AMENDIS Tanger	4,312		473							
	DU	AMENDIS Tetouan	2,292		299							
	DU	LYDEC	4,223		520			102	0.92	13.36		1
	VIU	ONEE	5,634	1,103	190	102		100	3.69	3.70	1.00	
	DU	RADEEL	2,953									
	DU	REDAL	3,759	1,773	442	186						
	DU	RAK	3,532		306							
	DU	RADEEMA	4,047		466				1.15			2
	DU	RADEM	2,750		301							
	DU	RADEEJ	4,048	1,615	436	168						1

table continues next page

Table E.4 Commercial Indicators (continued)

Country or economy	Utility type	Utility	Average consumption and billing				Collection rate	Metering Share of installed meters	Customer management and service quality			Duration of interruption
			Engy vol. sold per con.	Res. Engy vol. sold per con.	Billing per con.	Res. billing per con.			SAIFI	SAIDI	CAIDI	
			kWh/con.	kWh/con.	$/con.	$/con.	%	%	000	minutes	minutes	minutes
Oman	DU	RADEEF	2,814		312							
	DU	RADEES	2,621		302							
	GU	APBS	n.a.	n.a.	n.a.	n.a.	n.a.	n.a.	n.a.	n.a.	n.a.	n.a.
	GU	ABPC	n.a.	n.a.	n.a.	n.a.	n.a.	n.a.	n.a.	n.a.	n.a.	n.a.
	GU	ASPC	n.a.	n.a.	n.a.	n.a.	n.a.	n.a.	n.a.	n.a.	n.a.	n.a.
	GU	GPDCO	n.a.	n.a.	n.a.	n.a.	n.a.	n.a.	n.a.	n.a.	n.a.	n.a.
	GU	AKPP	n.a.	n.a.	n.a.	n.a.	n.a.	n.a.	n.a.	n.a.	n.a.	n.a.
	GU	ARPP	n.a.	n.a.	n.a.	n.a.	n.a.	n.a.	n.a.	n.a.	n.a.	n.a.
	GU	BPDP	n.a.	n.a.	n.a.	n.a.	n.a.	n.a.	n.a.	n.a.	n.a.	n.a.
	VIU	DPC	27,586	13,630		400						
	DU	MJEC				521						3
	DU	MZEC				432						
	DU	MEDC				582		15				3
	TU	OETC	n.a.	n.a.	n.a.	n.a.	n.a.	n.a.	n.a.	n.a.	n.a.	n.a.
	GU	PPC	n.a.	n.a.	n.a.	n.a.	n.a.	n.a.	n.a.	n.a.	n.a.	n.a.
	VIU	RAECO	23,011		925	459	71		1.75	1.88	n.a.	n.a.
	GU	SSPWC	n.a.	n.a.	n.a.	n.a.	n.a.	n.a.	n.a.	n.a.	n.a.	n.a.
	GU	SPP	n.a.	n.a.	n.a.	n.a.	n.a.	n.a.	n.a.	n.a.	n.a.	n.a.
	GU	UPC	n.a.	n.a.	n.a.	n.a.	n.a.	n.a.	n.a.	n.a.	n.a.	n.a.
	GU	WAJPCO	n.a.	n.a.	n.a.	n.a.	n.a.	n.a.	n.a.	n.a.	n.a.	n.a.
Qatar	VIU	KAHRAMAA	n.a.	n.a.	n.a.	n.a.	n.a.	n.a.	n.a.	n.a.	n.a.	n.a.

table continues next page

Table E.4 Commercial Indicators *(continued)*

			Commercial indicators									
			Average consumption and billing					Metering	Customer management and service quality			
			Engy vol. sold per con.	Res. Engy vol. sold per con.	Billing per con.	Res. billing per con.	Collection rate	Share of installed meters	SAIFI	SAIDI minutes	CAIDI minutes	Duration of interruption minutes
Country or economy	Utility type	Utility	kWh/con.	kWh/con.	$/con.	$/con.	%	%	000			
Saudi Arabia	VIU	SEC	35,937	22,154				100	4.09			
Tunisia	VIU	STEG	3,749		377			125				
West Bank	DU	JDECO	5,988	3,826	97	616						
West Bank	DU	NEDCO	4,307	2,476		504	90	100				
West Bank	DU	TUBAS	7,330	3,427		426	62	100				
Yemen, Rep.	VIU	PEC	2,631	1,922	178			100				

Source: MENA Electricity Database.

Note: CAIDI = Customer Average Interruption Duration Index; DU = distribution utility; GU = generation utility; kWh = kilowatt-hours; SAIDI = System Average Interruption Duration index; SAIFI = System Average Interruption Frequency Index; TU = transmission utility; VIU = vertically integrated utility.

233

Environmental Benefits Statement

The World Bank Group is committed to reducing its environmental footprint. In support of this commitment, we leverage electronic publishing options and print-on-demand technology, which is located in regional hubs worldwide. Together, these initiatives enable print runs to be lowered and shipping distances decreased, resulting in reduced paper consumption, chemical use, greenhouse gas emissions, and waste.

We follow the recommended standards for paper use set by the Green Press Initiative. The majority of our books are printed on Forest Stewardship Council (FSC)–certified paper, with nearly all containing 50–100 percent recycled content. The recycled fiber in our book paper is either unbleached or bleached using totally chlorine-free (TCF), processed chlorine–free (PCF), or enhanced elemental chlorine–free (EECF) processes.

More information about the Bank's environmental philosophy can be found at http://www.worldbank.org/corporateresponsibility.